STUDENT'S SOLUTIONS MANUAL

Basic Mathematics

6TH EDITION

Mervin L. Keedy
Purdue University

Marvin L. Bittinger
Indiana University -
Purdue University at Indianapolis

Judith A. Penna

ADDISON-WESLEY PUBLISHING COMPANY
Reading, Massachusetts • Menlo Park, California • New York
Don Mills, Ontario • Wokingham, England • Amsterdam • Bonn
Sydney • Singapore • Tokyo • Madrid • San Juan

Copyright © 1991 by Addison-Wesley Publishing Company, Inc.

ISBN 0-201-19666-2

6 7 8 9 10-CRS-97 96 95 94 93

TABLE OF CONTENTS

Special thanks are extended to Patsy Hammond for her
excellent typing and to Pam Smith for her careful
proofreading. Their patience, efficiency, and good
humor made the author's work much easier.

Exercise Set 1.1

1. 5742 = 5 thousands + 7 hundreds + 4 tens + 2 ones

3. 27,342 = 2 ten thousands + 7 thousands + 3 hundreds + 4 tens + 2 ones

5. 9010 = 9 thousands + 0 hundreds + 1 ten + 0 ones, or 9 thousands + 1 ten

7. 2300 = 2 thousands + 3 hundreds + 0 tens + 0 ones, or 2 thousands + 3 hundreds

9. 2 thousands + 4 hundreds + 7 tens + 5 ones = 2475

11. 6 ten thousands + 8 thousands + 9 hundreds + 3 tens + 9 ones = 68,939

13. 7 thousands + 3 hundreds + 0 tens + 4 ones = 7304

15. 1 thousand + 0 hundreds + 0 tens + 9 ones = 1009

17. 77 = seventy-seven

19.   8 8 , 0 0 0

Eighty-eight thousand

21.   1 2 3 , 7 6 5

One hundred twenty-three thousand, seven hundred sixty-five

23.   7 , 7 5 4 , 2 1 1

Seven million, seven hundred fifty-four thousand, two hundred eleven

25.   2 4 4 , 8 3 9 , 7 7 2

Two hundred forty-four million, eight hundred thirty-nine thousand, seven hundred seventy-two

27.   1 , 9 5 4 , 1 1 6

One million, nine hundred fifty-four thousand, one hundred sixteen

29. Two million, two hundred thirty-three thousand, eight hundred twelve

Standard notation is 2, 2 3 3, 8 1 2.

31. Eight billion

Standard notation is 8, 0 0 0 , 0 0 0 , 0 0 0.

33. Two hundred seventeen thousand, five hundred three

Standard notation is 2 1 7, 5 0 3.

35. Two million, one hundred seventy-three thousand, six hundred thirty-eight

Standard notation is 2, 1 7 3, 6 3 8.

37. Two hundred six million, six hundred fifty-eight thousand

Standard notation is 2 0 6, 6 5 8, 0 0 0.

39. 23 5 ,888

The digit 5 means 5 thousands.

41. 488, 5 26

The digit 5 means 5 hundreds.

43. 89, 3 02

The digit 3 tells the number of hundreds.

45. 89,3 0 2

The digit 0 tells the number of tens.

47. All digits are 9's. Answers may vary. For an 8-digit readout, for example, it would be 99,999,999.

Exercise Set 1.2

1. [She buys 3 yd one day.] [She buys 6 yd the next day.] [She bought 9 yd in all.]
   3 yd   +   6 yd   =   9 yd

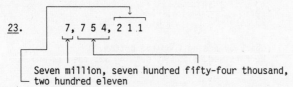

3. [Earns $23 one day] [Earns $31 the next day] [The total earned is $54]
   $23   + $31   = $54

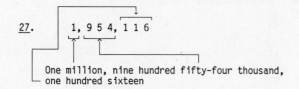

5.   3 6 4
    + 2 3   Add ones, add tens, then
     3 8 7   add hundreds.

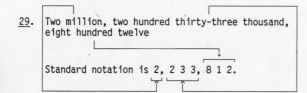

1

7. 1 7 1 6
   + 3 2 8 2
   ‾‾‾‾‾‾‾‾‾
     4 9 9 8

Add ones, add tens, add
hundreds, then add thousands

9.   ¹8 6
    + 7 8
   ‾‾‾‾‾‾‾
    1 6 4

Add ones: We get 14, so we
have 1 ten and 4 ones. Write
4 in the ones column and 1
above the tens. Add tens:
We get 16 tens.

11.   ¹9 ¹9 9
    +   1
   ‾‾‾‾‾‾‾‾‾‾‾
    1 0 0 0

Add ones: We get 10. Write
0 in the ones column and 1 above
the tens. Add tens: We get 10
tens. Write 0 in the tens column
and 1 above the hundreds. Add
hundreds: We get 10 hundreds.

13.   ¹9 ¹9 9
    + 1 1 1
   ‾‾‾‾‾‾‾‾‾
    1 1 1 0

Add ones: We get 10. Write 0
in the ones column and 1 above
the tens. Add tens: We get 11
tens. Write 1 in the tens column
and 1 above the hundreds. Add
hundreds: We get 11 hundreds.

15.   9 ¹0 9
    + 1 0 1
   ‾‾‾‾‾‾‾‾‾
    1 0 1 0

Add ones: We get 10. Write 0
in the ones column and 1 above
the tens. Add tens: We get 1 ten.
Add hundreds: We get 10 hundreds.

17.   ¹8 1 1
    + 3 9 0
   ‾‾‾‾‾‾‾‾‾
    1 2 0 1

Add ones: We get 1. Add tens:
We get 10 tens. Write 0 in the tens
column and 1 above the hundreds.
Add hundreds: We get 12 hundreds.

19.   ¹3 5 6
    + 4 9 1
   ‾‾‾‾‾‾‾‾‾
    8 4 7

Add ones: We get 7. Add tens:
We get 14 tens. Write 4 in the tens
column and 1 above the hundreds.
Add hundreds: We get 8 hundreds.

21.   ¹8 7 1 9
    + 1 4 2 0
   ‾‾‾‾‾‾‾‾‾‾‾
    1 0,1 3 9

Add ones: We get 9. Add tens:
We get 3 tens. Add hundreds: We
get 11 hundreds. Write 1 in the
hundreds column and 1 above the
thousands. Add thousands: We
get 10 thousands.

23.   ¹4 ¹8 2 5
    + 1 7 8 3
   ‾‾‾‾‾‾‾‾‾‾‾
    6 6 0 8

Add ones: We get 8. Add tens:
We get 10 tens. Write 0 in the
column and 1 above the hundreds.
Add hundreds: We get 16 hundreds.
Write 6 in the hundreds column and 1
above the thousands. Add thousands: We get 6
thousands.

25.   ¹9 ¹9 ¹9 9
    + 6 7 8 5
   ‾‾‾‾‾‾‾‾‾‾‾
    1 6,7 8 4

Add ones: We get 14. Write 4
in the ones column and 1 above
the tens. Add tens: We get
18 tens. Write 8 in the tens
column and 1 above the hundreds.
Add hundreds: We get 17 hundreds. Write 7 in the
hundreds column and 1 above the thousands. Add
thousands: We get 16 thousands.

27.   2 ¹3,¹4 ¹4 3
    + 1 0,9 8 9
   ‾‾‾‾‾‾‾‾‾‾‾‾‾
    3 4,4 3 2

Add ones: We get 12. Write
2 in the ones column and 1
above the tens. Add tens:
We get 13 tens. Write 3 in the
tens column and 1 above the
hundreds. Add hundreds: We get 14 hundreds. Write
4 in the hundreds column and 1 above the thousands.
Add thousands: We get 4 thousands. Add ten thousands:
We get 3 ten thousands.

29.   ¹7 ¹7,¹5 ¹4 3
    + 2 3,7 6 7
   ‾‾‾‾‾‾‾‾‾‾‾‾‾
    1 0 1,3 1 0

Add ones: We get 10. Write 0
in the ones column and 1 above
the tens. Add tens: We get 11
tens. Write 1 in the tens column
and 1 above the hundreds. Add
hundreds: We get 13 hundreds. Write 3 in the hundreds
column and 1 above the thousands. Add thousands: We
get 11 thousands. Write 1 in the thousands column and
1 above the ten thousands. Add ten thousands: We get
10 ten thousands.

31.   ¹9 ¹9,¹9 ¹9 9
    +   1 1 2
   ‾‾‾‾‾‾‾‾‾‾‾‾‾‾‾
    1 0 0,1 1 1

Add ones: We get 11. Write 1
in the ones column and 1 above
the tens. Add tens: We get 11
tens. Write 1 in the tens column
and 1 above the hundreds. Add
hundreds: We get 11 hundreds. Write 1 in the
hundreds column and 1 above the thousands. Add
thousands: We get 10 thousands. Write 0 in the
thousands column and 1 above the ten thousands. Add
ten thousands: We get 10 ten thousands.

33. Add from the top.
     We first add 7 and 9, getting 16; then 16 and 4,
     getting 20; then 20 and 8 getting 28.

     Check by adding from the bottom.
     We first add 8 and 4, getting 12; then 12 and 9,
     getting 21; then 21 and 7, getting 28.

35. Add from the top.

     Check:

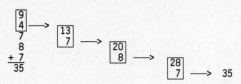 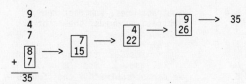 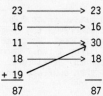 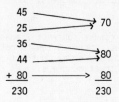

37. Add from the top.

```
 9
 4
 7
 8
+7
──
 35
```

9→13,7→20,8→28,7→35

Check:

```
 9
 4
 7
+8
 7
──
 35
```

39. We look for pairs of numbers whose sums are 10, 20, 30, and so on.

```
23 ───────→ 23
16 ───────→ 16
11 ──┐   ┌→ 30
18 ──┘   └→ 18
+19
───        ──
87         87
```

41. We look for pairs of numbers whose sums are 10, 20, 30, and so on.

```
45 ─┐
25 ─┴──→ 70
36 ─┐
44 ─┴──→ 80
+80 ──→ 80
───
230        230
```

43.
```
  1
  2 3
  6 2
+ 4 5
──────
1 3 0
```
Add ones: We get 10. Write 0 in the ones column and 1 above the tens. Add tens: We get 13 tens.

45.
```
  5 1
  3 6
+ 6 2
──────
1 4 9
```
Add ones, then add tens.

47.
```
  2 6
  8 2
+ 6 1
──────
1 6 9
```
Add ones, then add tens.

49.
```
  1 1
  2 0 7
  2 9 5
+ 3 4 0
────────
  8 4 2
```
Add ones: We get 12. Write 2 in the ones column and 1 above the tens. Add tens: We get 14 tens. Write 4 in the tens column and 1 above the hundreds. Add hundreds: We get 8 hundreds.

51.
```
  1 1 1
  2 0 3 7
  4 9 2 3
  3 4 7 1
+ 1 2 4 8
──────────
1 1,6 7 9
```
Add ones: We get 19. Write 9 in the ones column and 1 above the tens. Add tens: We get 17 tens. Write 7 in the tens column and 1 above the hundreds. Add hundreds: We get 16 hundreds. Write 6 in the hundreds column and 1 above the thousands. Add thousands: We get 11 thousands.

53.
```
    1   1
  3 4 2 0
  8 7 1 9
  4 3 1 2
+ 6 2 0 3
──────────
2 2,6 5 4
```
Add ones: We get 14. Write 4 in the ones column and 1 above the tens. Add tens: We get 5 tens. Add hundreds: We get 16 hundreds. Write 6 in the hundreds column and 1 above the thousands. Add thousands: We get 22 thousands.

55.
```
  2 2 3 3 1 1
  5,6 7 8,9 8 7
  1,4 0 9,3 1 2
    8 9 8,8 8 8
+ 4,7 7 7,9 1 0
────────────────
1 2,7 6 5,0 9 7
```

57. 7000 + 900 + 90 + 2 = 7 thousands + 9 hundreds + 9 tens + 2 ones = 7992

59. 1 + 99 = 100, 2 + 98 = 100,..., 49 + 51 = 100. Then 49 × 100 = 4900 and 4900 + 50 + 100 = 5050.

Exercise Set 1.3

1.
```
[Number of        [Number of      [Number of
 gallons to    -   gallons sold] =  gallons left]
 start with]
  2400 gal    -     800 gal    =   [        ]
```

3. 10 - 7 = 3
      ↑
   This number gets added (after 3).
      ↓
   10 = 3 + 7
   (By the commutative law of addition, 10 = 7 + 3 is also correct.)

5. 13 - 8 = 5
      ↑
   This number gets added (after 5).
      ↓
   13 = 5 + 8
   (By the commutative law of addition, 13 = 8 + 5 is also correct.)

7. 23 - 9 = 14
↑
This number gets
added (after 14).
↓
23 = 14 + 9

(By the commutative law of addition, 23 = 9 + 14
is also correct.)

9. 43 - 16 = 27
This number gets
added (after 27).
43 = 27 + 16

(By the commutative law of addition, 43 = 16 + 27
is also correct.)

11. 6 + 9 = 15   6 + 9 = 15
↑   ↑
This number gets   This number gets
subtracted (moved).   subtracted (moved).
↓   ↓
6 = 15 - 9   9 = 15 - 6

13. 8 + 7 = 15   8 + 7 = 15
↑   ↑
This number gets   This number gets
subtracted (moved).   subtracted (moved).
↓   ↓
8 = 15 - 7   7 = 15 - 8

15. 17 + 6 = 23   17 + 6 = 23
↑   ↑
This number gets   This number gets
subtracted (moved).   subtracted (moved).
↓   ↓
17 = 23 - 6   6 = 23 - 17

17. 23 + 9 = 32   23 + 9 = 32
↑   ↑
This number gets   This number gets
subtracted (moved).   subtracted (moved).
↓   ↓
23 = 32 - 9   9 = 32 - 23

19. We write an addition sentence first.

Number   Number yet   is   Sales
already sold   plus to be sold   goal

190   +   [   ]   =   220

Now we write a related subtraction.

190 + [   ] = 220

[   ] = 220 - 190   190 gets sub-
                      tracted (moved).

21.   1 6   Subtract ones, then subtract
    -   4   tens.
    ──────
      1 2

23.   6 5   Subtract ones, then subtract
    - 2 1   tens.
    ──────
      4 4

25.   8 6 6   Subtract ones, subtract tens,
    - 3 3 3   then subtract hundreds.
    ──────
      5 3 3

27.   4 5 4 7   Subtract ones, subtract tens,
    - 3 4 2 1   subtract hundreds, then subtract
    ──────   thousands.
      1 1 2 6

         7 16
29.   8̶ 6̶   We cannot subtract 7 ones from 6
    -   4 7   ones. Borrow 1 ten to get 16 ones.
    ──────   Subtract ones, then subtract tens.
      3 9

         11
        5̶ 1̶ 15
31.   6̶ 2̶ 5̶   We cannot subtract 7 ones from 5
    - 3 2 7   ones. Borrow 1 ten to get 15 ones.
    ──────   Subtract ones. We cannot subtract
      2 9 8   2 tens from 1 ten. Borrow 1
                 hundred to get 11 tens. Subtract
                 tens, then subtract hundreds.

         2 15
33.   8 3̶ 5̶   We cannot subtract 9 ones from 5
    - 6 0 9   ones. Borrow 1 ten to get 15 ones.
    ──────   Subtract ones, subtract tens, then
      2 2 6   subtract hundreds.

         7 11
34.   9 8̶ 1̶   We cannot subtract 7 ones from 1
35.   9 8̶ 1̶   one. Borrow 1 ten to get 11 ones.
    - 7 4 7   Subtract ones, subtract tens, then
    ──────   subtract hundreds.
      2 3 4

         6 16
37.   7 7̶ 8̶ 9   Subtract ones. We cannot
    - 2 3 8 7   subtract 8 tens from 6 tens.
    ──────   Borrow 1 hundred to get 16 tens.
      5 3 8 2   Subtract tens, subtract hundreds,
                 then subtract thousands.

           17
         8 7 12
39.   3 9̶ 8̶ 2̶   We cannot subtract 9 ones from 2
    - 2 4 8 9   ones. Borrow 1 ten to get 12 ones.
    ──────   Subtract ones. We cannot subtract
      1 4 9 3   8 tens from 7 tens. Borrow 1
                 hundred to get 17 tens. Subtract
                 tens, subtract hundreds, then
                 subtract thousands.

           13
         4 9 8 16
41.   5̶ 0̶ 4̶ 6̶   We cannot subtract 9 ones from 6
    - 2 8 5 9   ones. Borrow 1 ten to get 16 ones.
    ──────   Subtract ones. We cannot subtract
      2 1 8 7   5 tens from 3 tens. We have 5
                 thousands or 50 hundreds. We borrow
                 1 hundred to get 13 tens. We then
have 49 hundreds. Subtract tens, subtract hundreds,
then subtract thousands.

43.
$$
\begin{array}{r}
1513 \\
6\ 5\ 2\ 10 \\
\cancel{7}\ \cancel{6}\ \cancel{4}\ \cancel{0} \\
-\ 3\ 8\ 0\ 9 \\
\hline
3\ 8\ 3\ 1
\end{array}
$$

We cannot subtract 9 ones from 0 ones. Borrow 1 ten to get 10 ones. Subtract ones, then tens. We cannot subtract 8 hundreds from 6 hundreds. Borrow 1 thousand to get 16 hundreds. Subtract hundreds, then thousands.

45.
$$
\begin{array}{r}
11\ 15\ 13 \\
\cancel{1}\ \cancel{2}, \cancel{6}\ \cancel{4}\ 7 \\
-\ \ \ 4\ 8\ 9\ 9 \\
\hline
7\ 7\ 4\ 8
\end{array}
$$

47.
$$
\begin{array}{r}
16\ 16 \\
5\ \cancel{6}\ \cancel{6}\ 11 \\
4\ \cancel{6}, \cancel{7}\ \cancel{7}\ \cancel{1} \\
-\ 1\ 2, 9\ 7\ 7 \\
\hline
3\ 3, 7\ 9\ 4
\end{array}
$$

49.
$$
\begin{array}{r}
7\ 10 \\
\cancel{8}\ \cancel{0} \\
-\ 2\ 4 \\
\hline
5\ 6
\end{array}
$$

51.
$$
\begin{array}{r}
8\ 10 \\
\cancel{9}\ \cancel{0} \\
-\ 5\ 4 \\
\hline
3\ 6
\end{array}
$$

53.
$$
\begin{array}{r}
13 \\
3\ 10 \\
\cancel{1}\ \cancel{4}\ \cancel{0} \\
-\ \ \ 5\ 6 \\
\hline
8\ 4
\end{array}
$$

55.
$$
\begin{array}{r}
8\ 10 \\
6\ \cancel{9}\ \cancel{0} \\
-\ 2\ 3\ 6 \\
\hline
4\ 5\ 4
\end{array}
$$

57.
$$
\begin{array}{r}
8\ 10 \\
\cancel{9}\ \cancel{0}\ 3 \\
-\ 1\ 3\ 2 \\
\hline
7\ 7\ 1
\end{array}
$$

59.
$$
\begin{array}{r}
2\ 9\ 10 \\
2\ \cancel{3}\ \cancel{0}\ \cancel{0} \\
-\ \ \ 1\ 0\ 9 \\
\hline
2\ 1\ 9\ 1
\end{array}
$$

We have 3 hundreds or 30 tens. We borrow 1 ten to get 10 ones. We then have 29 tens. Subtract ones, then tens, then hundreds, and then thousands.

61.
$$
\begin{array}{r}
7\ 9\ 18 \\
6\ \cancel{8}\ \cancel{0}\ \cancel{8} \\
-\ 3\ 0\ 5\ 9 \\
\hline
3\ 7\ 4\ 9
\end{array}
$$

We have 8 hundreds or 80 tens. We borrow 1 ten to get 18 ones. We then have 79 tens. Subtract ones, then tens, then hundreds, and then thousands.

63.
$$
\begin{array}{r}
3\ 9\ 12 \\
4\ \cancel{0}\ \cancel{2}\ 7 \\
-\ 2\ 9\ 7\ 4 \\
\hline
1\ 0\ 5\ 3
\end{array}
$$

Subtract ones. We have 4 thousands or 40 hundreds. We borrow 1 ten to get 12 tens. We then have 39 hundreds. Subtract tens, then hundreds, then thousands.

65.
$$
\begin{array}{r}
6\ 9\ 9\ 10 \\
\cancel{7}\ \cancel{0}\ \cancel{0}\ \cancel{0} \\
-\ 2\ 7\ 9\ 4 \\
\hline
4\ 2\ 0\ 6
\end{array}
$$

We have 7 thousands or 700 tens. We borrow 1 ten to get 10 ones. We then have 699 tens. Subtract ones, then tens, then hundreds, then thousands.

67.
$$
\begin{array}{r}
7\ 9\ 9\ 10 \\
4\ \cancel{8}, \cancel{0}\ \cancel{0}\ \cancel{0} \\
-\ 3\ 7, 6\ 9\ 5 \\
\hline
1\ 0, 3\ 0\ 5
\end{array}
$$

We have 8 thousands or 800 tens. We borrow 1 ten to get 10 ones. We then have 799 tens. Subtract ones, then tens, then hundreds, then thousands, then ten thousands.

69. 6,3 $\boxed{7}$ 5,602

The digit 7 means 7 ten thousands.

Exercise Set 1.4

---

1. Round 48 to the nearest ten.

4 $\boxed{8}$

The digit 4 is in the tens place. Consider the next digit to the right. Since the digit, 8, is 5 or higher, round 4 tens up to 5 tens. Then change the digit to the right of the tens digit to zero.

The answer is 50.

3. Round 67 to the nearest ten.

6 $\boxed{7}$

The digit 6 is in the tens place. Consider the next digit to the right. Since the digit, 7, is 5 or higher, round 6 tens up to 7 tens. Then change the digit to the right of the tens digit to zero.

The answer is 70.

5. Round 731 to the nearest ten.

73 $\boxed{1}$

The digit 3 is in the tens place. Consider the next digit to the right. Since the digit, 1, is 4 or lower, round down, meaning that 3 tens stays as 3 tens. Then change the digit to the right of the tens digit to zero.

The answer is 730.

7. Round 895 to the nearest ten.

89 $\boxed{5}$

The digit 9 is in the tens place. Consider the next digit to the right. Since the digit, 5, is 5 or higher, we round up. The 89 tens become 90 tens. Then change the digit to the right of the tens digit to zero.

The answer is 900.

9. Round 146 to the nearest hundred.

1 $\boxed{4}$ 6

The digit 1 is in the hundreds place. Consider the next digit to the right. Since the digit, 4, is 4 or lower, round down, meaning that 1 hundred stays as 1 hundred. Then change all digits to the right of the hundreds digit to zeros.

The answer is 100.

11. Round 957 to the nearest hundred.

9⑤7
↑

The digit 9 is in the hundreds place. Consider the next digit to the right. Since the digit, 5, is 5 or higher, round up. The 9 hundreds become 10 hundreds. Then change all digits to the right of the hundreds digit to zeros.

The answer is 1000.

13. Round 3583 to the nearest hundred.

35⑧3
↑

The digit 5 is in the hundreds place. Consider the next digit to the right. Since the digit, 8, is 5 or higher, round 5 hundreds up to 6 hundreds. Then change all digits to the right of the hundreds digit to zeros.

The answer is 3600.

15. Round 2850 to the nearest hundred.

28⑤0
↑

The digit 8 is in the hundreds place. Consider the next digit to the right. Since the digit, 5, is 5 or higher, round 8 hundreds up to 9 hundreds. Then change all digits to the right of the hundreds digit to zeros.

The answer is 2900.

17. Round 5932 to the nearest thousand.

5⑨32
↑

The digit 5 is in the thousands place. Consider the next digit to the right. Since the digit, 9, is 5 or higher, round 5 thousands up to 6 thousands. Then change all digits to the right of the thousands digit to zeros.

The answer is 6000.

19. Round 7500 to the nearest thousand.

7⑤00
↑

The digit 7 is in the thousands place. Consider the next digit to the right. Since the digit, 5, is 5 or higher, round 7 thousands up to 8 thousands. Then change all digits to the right of the thousands digit to zeros.

The answer is 8000.

21. Round 45,340 to the nearest thousand.

45,③40
↑

The digit 5 is in the thousands place. Consider the next digit to the right. Since the digit, 3, is 4 or lower, round down, meaning that 5 thousands stays as 5 thousands. Then change all digits to the right of the thousands digit to zeros.

The answer is 45,000.

23. Round 373,405 to the nearest thousand.

373,④05
↑

The digit 3 is in the thousands place. Consider the next digit to the right. Since the digit, 4, is 4 or lower, round down, meaning that 3 thousands stays as 3 thousands. Then change all digits to the right of the thousands digit to zeros.

The answer is 373,000.

25.   Rounded to the nearest ten

7848   7850
+9747   + 9750
                     1 7,6 0 0 ←— Estimated answer

27.   Rounded to the nearest ten

6882   6880
-1748   - 1750
                     5 1 3 0 ←— Estimated answer

29.   Rounded to the nearest ten

  45   50
  77   80
  25   30
+ 56   + 60
 343   2 2 0 ←— Estimated answer

The sum 343 seems to be incorrect since 220 is not close to 343.

31.   Rounded to the nearest ten

 622   620
  78   80
  81   80
+111   +110
 932   8 9 0 ←— Estimated answer

The sum 932 seems to be incorrect since 890 is not close to 932.

33.   Rounded to the nearest hundred

7848   7800
+9747   + 9700
                     1 7,5 0 0 ←— Estimated answer

35.   Rounded to the nearest hundred

6852   6900
-1748   - 1700
                     5 2 0 0 ←— Estimated answer

37.   Rounded to the nearest hundred

```
    2 1 6              2 0 0
      8 4              1 0 0
    7 4 5              7 0 0
  + 5 9 5            + 6 0 0
    1 6 4 0            1 6 0 0  ← Estimated answer
```

The sum 1640 seems to be correct since 1600 is close to 1640.

39.   Rounded to the nearest hundred

```
    7 5 0              8 0 0
    4 2 8              4 0 0
      6 3              1 0 0
  + 2 0 5            + 2 0 0
    1 4 4 6            1 5 0 0  ← Estimated answer
```

The sum 1446 seems to be correct since 1500 is close to 1446.

41.   Rounded to the nearest thousand

```
    9 6 4 3            1 0,0 0 0
    4 8 2 1            5 0 0 0
    8 9 4 3            9 0 0 0
  + 7 0 0 4          + 7 0 0 0
                      3 1,0 0 0  ← Estimated answer
```

43.   Rounded to the nearest thousand

```
    9 2,1 4 9          9 2,0 0 0
  − 2 2,5 5 5        − 2 3,0 0 0
                      6 9,0 0 0  ← Estimated answer
```

45.
```
  —+——+——+———++——→
   0   5  10  17
```
Since 0 is to the left of 17, 0 < 17.

47.
```
  —+———++—+———+———+———+——++——→
   5    12 15   20   25  30 34
```
Since 34 is to the right of 12, 34 > 12.

49.
```
  —+———+————————→
 1000 1001
```
Since 1000 is to the left of 1001, 1000 < 1001.

51.
```
  —+———+————————→
  132 133
```
Since 133 is to the right of 132, 133 > 132.

53.
```
  —++—+————+———+———+————+——+———→
  0 17   100  200  300   400 460 500
```
Since 460 is to the right of 17, 460 > 17.

55.
```
  —+———++——+———+———+———++——→
   0    11  20   30   37
```
Since 37 is to the right of 11, 37 > 11.

57.
```
  1 1 1 1
  6 7,7 8 9
+ 1 8,9 6 5
  8 6,7 5 4
```
Add ones: We get 14. Write 4 in the ones column and 1 above the tens. Add tens: We get 15 tens. Write 5 in the tens column and 1 above the hundreds. Add hundreds: We get 17 hundreds. Write 7 in the hundreds column and 1 above the thousands. Add thousands: We get 16 thousands. Write 6 in the thousands column and 1 above the ten thousands. Add ten thousands: We get 8 ten thousands.

59. Using a calculator, we find that the sum is 30,411.

61. Using a calculator, we find that the difference is 69,594.

Exercise Set 1.5

1. Repeated addition fits best in this case.

   $\boxed{\$10}$ $\boxed{\$10}$ $\boxed{\$10}$ · · · $\boxed{\$10}$

   32 addends

   32 · $10 = $320

3. We have a rectangular array.

   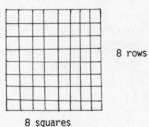

   8 rows

   8 squares

   8 · 8 = 64

5. If we think of filling the rectangle with square feet, we have a rectangular array.

   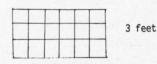

   3 feet

   6 feet

   A = ℓ × w = 6 × 3 = 18 ft^2

7.
```
    8 7
  × 1 0
    8 7 0
```
Multiplying by 1 ten (We write 0 and then multiply 87 by 1.)

9.
```
       2 3 4 0
     × 1 0 0 0
  2,3 4 0,0 0 0
```
Multiplying by 1 thousand (We write 000 and then multiply 2340 by 1.)

11.
```
    4
    6 5
  ×   8
  5 2 0      Multiplying by 8
```

13.
```
    2
    9 4
  ×   6
  5 6 4      Multiplying by 6
```

15.
```
    0
  5   9
  ×     3
  1 5 2 7    Multiplying by 3
```

17.
```
    1 2 6
  9 2 2 9
  ×       7
  6 4,6 0 3  Multiplying by 7
```

19.
```
    2
    5 3
  ×   9 0
  4 7 7 0    Multiplying by 9 tens (We
             write 0 and then multiply
             53 by 9.)
```

21.
```
    2
    4
    6 5
  ×   4 8
  5 2 0      Multiplying by 8
  2 6 0 0    Multiplying by 40
  3 1 2 0    Adding
```

23.
```
    2
    6 4 0
  ×     7 2
  1 2 8 0    Multiplying by 2
  4 4 8 0 0  Multiplying by 70
  4 6,0 8 0  Adding
```

25.
```
    1 1
    4 4 4
  ×     3 3
  1 3 3 2    Multiplying by 3
  1 3 3 2 0  Multiplying by 30
  1 4,6 5 2  Adding
```

27.
```
    3
  5   9
  ×   4 0 8
  4 0 7 2   ◄─  Multiplying by 8
                Multiplying by 4 hundreds (We
  2 0 3 6 0 0 ◄─ write 00 and then multiply 509
                by 4.)
  2 0 7,6 7 2    Adding
```

29.
```
    5 5
    3 3
    6 7 8
  ×   7 4 2
    1 3 5 6    Multiplying by 2
  2 7 1 2 0    Multiplying by 40
  4 7 4 6 0 0  Multiplying by 700
  5 0 3,0 7 6  Adding
```

31.
```
    2 2
    3 3
    4 8 9          Multiplying by 4 tens (We
  ×   3 4 0        write 0 and then multiply 489
                   by 4.)
    1 9 5 6 0  ◄─  Multiplying by 3 hundreds (We
  1 4 6,7 0 0 ◄─   write 00 and then multiply 489
  1 6 6,2 6 0      by 3.)
                   Adding
```

33.
```
    2 1 1
    5 7 7
    1 3 3
    4 3 7 8
  ×   2 6 9 4
    1 7 5 1 2      Multiplying by 4
    3 9 4 0 2 0    Multiplying by 90
  2 6 2 6 8 0 0    Multiplying by 600
  8 7 5 6 0 0 0    Multiplying by 2000
  1 1,7 9 4,3 3 2  Adding
```

35.
```
    2
    1 1 1
    1 1 1
    6 4 2 8
  ×   3 2 2 4
    2 5 7 1 2      Multiplying by 4
    1 2 8 5 6 0    Multiplying by 20
  1 2 8 5 6 0 0    Multiplying by 200
  1 9 2 8 4 0 0 0  Multiplying by 3000
  2 0,7 2 3,8 7 2  Adding
```

37.
```
    1 3
    3 4 8 2
  ×     1 0 4      Multiplying by 4
  1 3 9 2 8   ◄─  Multiplying by 1 hundred (We
                   write 00 and then multiply 3482
  3 4 8 2 0 0 ◄─   by 1.)
  3 6 2,1 2 8      Adding
```

39.
```
    2
    4
    5 0 0 6
  ×   4 0 0 8      Multiplying by 8
  4 0 0 4 8   ◄─  Multiplying by 4 thousands (We
                   write 000 and then multiply
  2 0 0 2 4 0 0 0 ◄─ 5006 by 4.)
  2 0,0 6 4,0 4 8  Adding
```

41.
```
    2 3
    2 3
    5 6 0 8
  ×   4 5 0 0
  2 8 0 4 0 0 0    Multiplying by 500 (We write
                   00 and then multiply 5608 by 5.)
  2 2 4 3 2 0 0 0  Multiplying by 4000
  2 5,2 3 6,0 0 0  Adding
```

43.
```
  2 1
  3 3
  8 7 6
× 3 4 5
─────────
  4 3 8 0      Multiplying by 5
3 5 0 4 0      Multiplying by 40
2 6 2 8 0 0    Multiplying by 300
─────────
3 0 2,2 2 0    Adding
```

45.
```
  5 5
  1 1
  1 1
  7 8 8 9
× 6 2 2 4
───────────
    3 1 5 5 6    Multiplying by 4
  1 5 7 7 8 0    Multiplying by 20
1 5 7 7 8 0 0    Multiplying by 200
4 7 3 3 4 0 0 0  Multiplying by 6000
───────────
4 9,1 0 1,1 3 6  Adding
```

47.
```
  2 2
  5 5 5
×   5 5
─────────
  2 7 7 5      Multiplying by 5
2 7 7 5 0      Multiplying by 50
─────────
3 0,5 2 5      Adding
```

49.
```
  1 1
  7 3 4
× 4 0 7
─────────
    5 1 3 8 ←── Multiplying by 7
2 9 3 6 0 0 ←── Multiplying by 4 hundreds (We
─────────        write 00 and then multiply 734
2 9 8,7 3 8      by 4.)
                 Adding
```

51.   Rounded to the nearest ten
```
  4 5                      5 0
× 6 7                    × 7 0
                     3 5 0 0 ←── Estimated answer
```

53.   Rounded to the nearest ten
```
  3 4                      3 0
× 2 9                    × 3 0
                       9 0 0 ←── Estimated answer
```

55.   Rounded to the nearest hundred
```
  8 7 6                    9 0 0
× 3 4 5                  × 3 0 0
                   2 7 0,0 0 0 ←── Estimated answer
```

57.   Rounded to the nearest hundred
```
  4 3 2                    4 0 0
× 1 9 9                  × 2 0 0
                    8 0,0 0 0 ←── Estimated answer
```

59.   Rounded to the nearest thousand
```
  5 6 0 8                  6 0 0 0
× 4 5 7 6                × 5 0 0 0
                 3 0,0 0 0,0 0 0 ←── Estimated answer
```

61.   Rounded to the nearest thousand
```
  7 8 8 8                  8 0 0 0
× 6 2 2 4                × 6 0 0 0
                  4 8,0 0 0,0 0 0 ←── Estimated answer
```

63.
```
  1
    2 0        Add ones: We get 0.  Add tens.
  8 5 0        We get 7 tens.  Add hundreds.  We
+ 3 5 0 0      get 13 hundreds.  Write 3 in the
─────────      hundreds column and 1 above the
  4 3 7 0      thousands.  Add thousands:  We
               get 4 thousands.
```

65. Round 234⬚5 to the nearest ten.

The digit 4 is in the tens place. Since the next digit to the right, 5, is 5 or higher, round 4 tens up to 5 tens. Then change the digit to the right of the tens digit to zero.

The answer is 2350.

Round 23⬚45 to the nearest hundred.

The digit 3 is in the hundreds place. Since the next digit to the right, 4, is 4 or lower, round down, meaning 3 hundreds stays as 3 hundreds. Then change all digits to the right of the hundreds digit to zeros.

The answer is 2300.

Round 2⬚345 to the nearest thousand.

The digit 2 is in the thousands place. Since the next digit to the right, 3, is 4 or lower, round down, meaning 2 thousands stays as 2 thousands. Then change all digits to the right of the thousands digit to zeros.

The answer is 2000.

Exercise Set 1.6

1. Think of an array with 4 lb in each row. The candy in each row will go in a box.

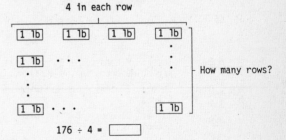

$176 \div 4 = \boxed{}$

3. Think of an array with $23,000 in each row. The money in each row will go to a child.

23,000 in each row

| $1 | | $1 | | · · · | | $1 |

| $1 | | · · ·

· ← How many rows?

·

·

| $1 | | · · · | | $1 |

$184,000 ÷ $23,000 = [   ]

5. 24 ÷ 8 = 3   The 8 moves to the right. A related multiplication sentence is 24 = 3·8.

7. 22 ÷ 22 = 1   The 22 on the right of the ÷ symbol moves to the right. A related multiplication sentence is 22 = 1·22.

9. 54 ÷ 6 = 9   The 6 moves to the right. A related multiplication sentence is 54 = 9·6.

11. 37 ÷ 1 = 37   The 1 moves to the right. A related multiplication sentence is 37 = 37·1.

13. 9 × 5 = 45

Move a factor to the other side and then write a division.

9 × 5 = 45   9 × 5 = 45

9 = 45 ÷ 5   5 = 45 ÷ 9

15. Two related division sentences for 37·1 = 37 are:

37 = 37 ÷ 1   (37·1 = 37)

and

1 = 37 ÷ 37   (37·1 = 37)

17. 8 × 8 = 64

Since the factors are both 8, moving either one to the other side gives the related division sentence 8 = 64 ÷ 8.

19. Two related division sentences for 11·6 = 66 are:

11 = 66 ÷ 6   (11·6 = 66)

and

6 = 66 ÷ 11   (11·6 = 66)

21.
```
      55
  5)2 7 7
    2 5 0
      2 7
      2 5
        2
```
Think: 2 hundreds ÷ 5. There are no hundreds in the quotient.

Think: 27 tens ÷ 5. Estimate 5 tens.

Think: 27 ones ÷ 5. Estimate 5 ones.

The answer is 55 R 2.

23.
```
    1 0 8
  8)8 6 4
    8 0 0
      6 4
      6 4
        0
```
Think: 8 hundreds ÷ 8. Estimate 1 hundred.

Think: 6 tens ÷ 8. There are no tens in the quotient (other than the tens in 100). Write a 0 to show this.

Think: 64 ones ÷ 8. Estimate 8 ones.

The answer is 108.

25.
```
      3 0 7
  4)1 2 2 8
    1 2 0 0
        2 8
        2 8
          0
```
Think: 12 hundreds ÷ 4. Estimate 3 hundreds.

Think: 2 tens ÷ 4. There are no tens in the quotient (other than the tens in 300). Write a 0 to show this.

Think: 28 ones ÷ 4. Estimate 7 ones.

The answer is 307.

27.
```
      9 0 6
  7)6 3 4 5
    6 3 0 0
        4 5
        4 2
          3
```
Think: 63 hundreds ÷ 7. Estimate 9 hundreds.

Think: 4 tens ÷ 7. There are no tens in the quotient other than the tens in 900). Write a 0 to show this.

Think: 45 ones ÷ 7. Estimate 6 ones.

The answer is 906 R 3.

29.
```
      7 4
  4)2 9 7
    2 8 0
      1 7
      1 6
        1
```
Think: 29 tens ÷ 4. Estimate 7 tens.

Think: 17 ones ÷ 4. Estimate 4 ones.

The answer is 74 R 1.

31.
```
      9 2
  8)7 3 8
    7 2 0
      1 8
      1 6
        2
```
Think: 73 tens ÷ 8. Estimate 9 tens.

Think: 18 ones ÷ 8. Estimate 2 ones.

The answer is 92 R 2.

33.
```
      1 7 0 3
  5)8 5 1 5
    5 0 0 0
    3 5 1 5
    3 5 0 0
        1 5
        1 5
          0
```
Think: 8 thousands ÷ 5. Estimate 1 thousand.

Think: 35 hundreds ÷ 5. Estimate 7 hundreds.

Think: 1 ten ÷ 5. There are no tens in the quotient (other than the tens in 1700). Write a 0 to show this.

Think: 15 ones ÷ 5. Estimate 3 ones.

The answer is 1703.

35.
```
      9 8 7
9│8 8 8 8
      8 1 0 0
        7 8 8
          7 2 0
            6 8
            6 3
              5
```
Think: 88 hundreds ÷ 9.
Estimate 9 hundreds.

Think: 78 tens ÷ 9. Estimate 8 tens.

Think: 68 ones ÷ 9. Estimate 7 ones.

The answer is 987 R 5.

37.
```
        5 2
7 0│3 6 9 2
    3 5 0 0
      1 9 2
      1 4 0
        5 2
```
Think: 369 tens ÷ 70. Estimate 5 tens.

Think: 192 ones ÷ 70. Estimate 2 ones.

The answer is 52 R 52.

39.
```
        2 9
3 0│8 7 5
    6 0 0
    2 7 5
    2 7 0
        5
```
Think: 87 tens ÷ 30. Estimate 2 tens.

Think: 275 ones ÷ 30. Estimate 9 ones.

The answer is 29 R 5.

41.
```
        4 0
2 1│8 5 2
    8 4 0
      1 2
```
Round 21 to 20

Think: 85 tens ÷ 20. Estimate 4 tens.

Think: 12 ones ÷ 20. There are no ones in the quotient (other than the ones in 40). Write a 0 to show this.

The answer is 40 R 12.

43.
```
            8
8 5│7 6 7 2
    6 8 0 0
      8 7 2
```
Round 85 to 90.

Think: 767 tens ÷ 90. Estimate 8 tens.

Since 87 is larger than the divisor, the estimate is too low.

```
          9 0
8 5│7 6 7 2
    7 6 5 0
        2 2
```
Think: 767 tens ÷ 90. Estimate 9 tens.

Think: 22 ones ÷ 90. There are no ones in the quotient (other than the ones in 90). Write a 0 to show this.

The answer is 90 R 22.

45.
```
              3
1 1 1│3 2 1 9
      3 3 3 0
```
Round 111 to 100.

Think: 321 tens ÷ 100. Estimate 3 tens.

Since we cannot subtract 3330 from 3219, the estimate is too high.

```
            2 9
1 1 1│3 2 1 9
      2 2 2 0
          9 9 9
          9 9 9
              0
```
Think: 321 tens ÷ 100. Estimate 2 tens.

Think: 999 ones ÷ 100. Estimate 9 ones.

The answer is 29.

47.
```
        1 0 5
8│8 4 3
  8 0 0
      4 3
      4 0
        3
```
Think: 8 hundreds ÷ 8. Estimate 1 hundred.

Think: 4 tens ÷ 8. There are no tens in the quotient (other than the tens in 100). Write a 0 to show this.

Think: 43 ones ÷ 8. Estimate 5 ones.

The answer is 105 R 3.

49.
```
        5 0 7
6│3 0 4 3
  3 0 0 0
        4 3
        4 2
          1
```
Think: 30 hundreds ÷ 6. Estimate 5 hundreds.

Think: 4 tens ÷ 6. There are no tens in the quotient (other than the tens in 500). Write a 0 to show this.

Think: 43 ones ÷ 6. Estimate 7 ones.

The answer is 507 R 1.

51.
```
        1 0 0 7
5│5 0 3 6
  5 0 0 0
        3 6
        3 5
          1
```
Think: 5 thousands ÷ 5. Estimate 1 thousand.

Think: 0 hundreds ÷ 5. There are no hundreds in the quotient (other than the hundreds in 1000). Write a 0 to show this.

Think: 3 tens ÷ 5. There are no tens in the quotient (other than the tens in 1000). Write a 0 to show this.

Think: 36 ones ÷ 5. Estimate 7 ones.

The answer is 1007 R 1.

53.
```
        2 2
4 6)1 0 5 8
    9 2 0
    1 3 8
      9 2
      4 6
```
Round 46 to 50.

Think: 105 tens ÷ 50. Estimate 2 tens.

Think: 138 ones ÷ 50. Estimate 2 ones.

Since 46 is not smaller than the divisor, 46, the estimate is too low.

```
        2 3
4 6)1 0 5 8
    9 2 0
    1 3 8
    1 3 8
        0
```
Think: 138 ones ÷ 50. Estimate 3 ones.

The answer is 23.

55.
```
        1 0 7
3 2)3 4 2 5
    3 2 0 0
      2 2 5
      2 2 4
          1
```
Round 32 to 30.

Think: 34 hundreds ÷ 30. Estimate 1 hundred.

Think: 22 tens ÷ 30. There are no tens in the quotient. Write 0 to show this.

Think: 225 ones ÷ 30. Estimate 7 ones.

The answer is 107 R 1.

57.
```
        4
2 4)8 8 8 0
    9 6 0 0
```
Round 24 to 20.

Think: 88 hundreds ÷ 20. Estimate 4 hundreds.

Since we cannot subtract 9600 from 8880, the estimate is too high.

```
        3 8
2 4)8 8 8 0
    7 2 0 0
    1 6 8 0
    1 9 2 0
```
Think: 88 hundreds ÷ 20. Estimate 3 hundreds.

Think: 168 tens ÷ 20. Estimate 8 tens.

Since we cannot subtract 1920 from 1680, the estimate is too high.

```
        3 7 0
2 4)8 8 8 0
    7 2 0 0
    1 6 8 0
    1 6 8 0
          0
```
Think: 168 tens ÷ 20. Estimate 7 tens.

Think: 0 ones ÷ 20. There are no ones in the quotient (other than the ones in 370). Write a 0 to show this.

The answer is 370.

59.
```
          5
2 8)1 7,0 6 7
    1 4 0 0 0
    [3 0]6 7
```
Round 28 to 30.

Think: 170 hundreds ÷ 30. Estimate 5 hundreds.

Since 30 is larger than the divisor, 28, the estimate is too low.

```
          6 0 8
2 8)1 7,0 6 7
    1 6 8 0 0
        2 6 7
        2 2 4
        [4 3]
```
Think: 170 hundreds ÷ 30. Estimate 6 hundreds.

Think: 26 tens ÷ 30. There are no tens in the quotient other than the tens in 600). Write a 0 to show this.

Think: 267 ones ÷ 30. Estimate 8 ones.

Since 43 is larger than the divisor, 28, the estimate is too low.

```
          6 0 9
2 8)1 7,0 6 7
    1 6 8 0 0
        2 6 7
        2 5 2
          1 5
```
Think: 267 ones ÷ 30. Estimate 9 ones.

The answer is 609 R 15.

61.
```
          3 0 4
8 0)2 4,3 2 0
    2 4 0 0 0
        3 2 0
        3 2 0
            0
```
Think: 243 hundreds ÷ 80. Estimate 3 hundreds.

Think: 32 tens ÷ 80. There are no tens in the quotient. Write a 0 to show this.

Think: 320 ones ÷ 80. Estimate 4 ones.

The answer is 304.

63.
```
            3 5 0 8
2 8 5)9 9 9,9 9 9
      8 5 5 0 0 0
      1 4 4 9 9 9
      1 4 2 5 0 0
          2 4 9 9
          2 2 8 0
            2 1 9
```
The answer is 3508 R 219.

65.
```
              8 0 7 0
4 5 6)3,6 7 9,9 2 0
      3 6 4 8 0 0 0
          3 1 9 2 0
          3 1 9 2 0
                  0
```
The answer is 8070.

67. 7882 = 7 thousands + 8 hundreds + 8 tens + 2 ones

<u>69</u>. First we divide to find how many new cigarettes can be made from 29 cigarette butts:

$$\begin{array}{r} 7 \\ 4\overline{)2\,9} \\ \underline{2\,8} \\ 1 \end{array}$$

The person can make 7 new cigarettes, and there is 1 butt left over.

We divide again to find how many new cigarettes can be made by reusing the butts from the first seven cigarettes and the butt left over when the first seven were made. That is, the person now has a total of 8 butts to use:

$$\begin{array}{r} 2 \\ 4\overline{)8} \\ \underline{8} \\ 0 \end{array}$$

The person can now make 2 more cigarettes, so 7 + 2, or 9, cigarettes were made in all. After the last 2 cigarettes are smoked, 2 butts will be left over.

Exercise Set 1.7

<u>1</u>. x + 0 = 14

We replace x by different numbers until we get a true equation. If we replace x by 14, we get a true equation: 14 + 0 = 14. No other replacement makes the equation true, so the solution is 14.

<u>3</u>. y·17 = 0

We replace y by different numbers until we get a true equation. If we replace y by 0 we get a true equation: 0·17 = 0. No other replacement makes the equation true, so the solution is 0.

<u>5</u>.   13 + x = 42

13 + x - 13 = 42 - 13   Subtracting 13 on both sides

0 + x = 29   13 plus x minus 13 is 0 + x.

x = 29

<u>7</u>.   12 = 12 + m

12 - 12 = 12 + m - 12   Subtracting 12 on both sides

0 = 0 + m   12 plus m minus 12 is 0 + m.

0 = m

<u>9</u>. 3·x = 24

$\dfrac{3 \cdot x}{3} = \dfrac{24}{3}$ Dividing by 3 on both sides

x = 8   3 times x divided by 3 is x.

<u>11</u>. 112 = n·8

$\dfrac{112}{8} = \dfrac{n \cdot 8}{8}$ Dividing by 8 on both sides

14 = n

<u>13</u>. 45 × 23 = x

To solve the equation we carry out the calculation.

$$\begin{array}{r} 4\,5 \\ \times\ \ 2\,3 \\ \hline 1\,3\,5 \\ 9\,0\ \ \\ \hline 1\,0\,3\,5 \end{array}$$

The solution is 1035.

<u>15</u>. t = 125 ÷ 5

To solve the equation we carry out the calculation.

$$\begin{array}{r} 2\,5 \\ 5\overline{)1\,2\,5} \\ \underline{1\,0\,0} \\ 2\,5 \\ \underline{2\,5} \\ 0 \end{array}$$

The solution is 25.

<u>17</u>. p = 908 - 458

To solve the equation we carry out the calculation.

$$\begin{array}{r} 9\,0\,8 \\ -\ 4\,5\,8 \\ \hline 4\,5\,0 \end{array}$$

The solution is 450.

<u>19</u>. x = 12,345 + 78,555

To solve the equation we carry out the calculation.

$$\begin{array}{r} 1\,2{,}3\,4\,5 \\ +\ 7\,8{,}5\,5\,5 \\ \hline 9\,0{,}9\,0\,0 \end{array}$$

The solution is 90,900.

<u>21</u>. 3·m = 96

$\dfrac{3 \cdot m}{3} = \dfrac{96}{3}$ Dividing by 3 on both sides

m = 32

<u>23</u>. 715 = 5·z

$\dfrac{715}{5} = \dfrac{5 \cdot z}{5}$ Dividing by 5 on both sides

143 = z

<u>25</u>.   10 + x = 89   <u>27</u>.   61 = 16 + y

10 + x - 10 = 89 - 10   61 - 16 = 16 + y - 16

x = 79   45 = y

29. 6·p = 1944

$\dfrac{6 \cdot p}{6} = \dfrac{1944}{6}$

p = 324

31. 5·x = 3715

$\dfrac{5 \cdot x}{5} = \dfrac{3715}{5}$

x = 743

33.   47 + n = 84

47 + n - 47 = 84 - 47

n = 37

35.   x + 78 = 144

x + 78 - 78 = 144 - 78

x = 66

37. 165 = 11·n

$\dfrac{165}{11} = \dfrac{11 \cdot n}{11}$

15 = n

39. 624 = t·13

$\dfrac{624}{13} = \dfrac{t \cdot 13}{13}$

48 = t

41.   x + 214 = 389

x + 214 - 214 = 389 - 214

x = 175

43.   567 + x = 902

567 + x - 567 = 902 - 567

x = 335

45. 18·x = 1872

$\dfrac{18 \cdot x}{18} = \dfrac{1872}{18}$

x = 104

47. 40·x = 1800

$\dfrac{40 \cdot x}{40} = \dfrac{1800}{40}$

x = 45

49.   2344 + y = 6400

2344 + y - 2344 = 6400 - 2344

y = 4056

51. 8322 + 9281 = x

17,603 = x   Doing the addition

53. 234 × 78 = y

18,252 = y   Doing the multiplication

55. 58·m = 11,890

$\dfrac{58 \cdot m}{58} = \dfrac{11,890}{58}$

m = 205

57. x·198 = 10,890

$\dfrac{x \cdot 198}{198} = \dfrac{10,890}{198}$

x = 55

59. Two related division sentences for 6·8 = 48 are:

6 = 48 ÷ 8   (6·8 = 48 ↑)

and

8 = 48 ÷ 6   (6·8 = 48 ↑)

61. ──┼──┼──┼──┼──────→
      339 340 341 342

Since 342 is to the right of 339, 342 > 339.

Exercise Set 1.8

1. Familiarize. We visualize the situation.

Ty Cobb's singles + Stan Musial's singles = Total singles

3052 singles   2641 singles   n singles

Translate. We write a number sentence that corresponds to the situation:

3052 + 2641 = n

Solve. We carry out the addition.

    3 0 5 2
  + 2 6 4 1
    5 6 9 3

Check. We can repeat the calculation. We can also find an estimated answer by rounding: 3052 + 2641 ≈ 3000 + 2600 = 5600 ≈ 5693. Since the estimated answer is close to the calculation, our answer seems correct.

State. Ty Cobb and Stan Musial hit 5693 singles together.

3. Familiarize. We first make a drawing.

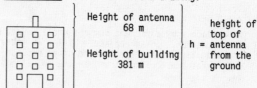

Since we are combining lengths, addition can be used. We let h = the height of the top of the antenna from the ground.

Translate. We translate to the following addition sentence:

381 + 68 = h

Solve. To solve we carry out the addition.

    3 8 1
  +   6 8
    4 4 9

Check. We can repeat the calculation. We can also find an estimated answer by rounding: 381 + 68 ≈ 380 + 70 = 450 ≈ 449. The answer checks.

State. The top of the antenna is 449 m from the ground.

5. <u>Familiarize</u>. We first make a drawing.

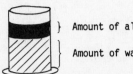
Amount of alcohol
Amount of water
$\left.\begin{array}{l}\\\\\end{array}\right\}$ a = Total amount of liquid

Since we are combining amounts, addition can be used. We let a = the total amount of liquid.

<u>Translate</u>. We translate to the following addition sentence:

$$2340 + 655 = a$$

<u>Solve</u>. To solve we carry out the addition.

$$\begin{array}{r} 2340 \\ +\ 655 \\ \hline 2995 \end{array}$$

<u>Check</u>. We can repeat the calculation. We can also find an estimated answer by rounding: 2340 + 655 ≈ 2300 + 700 = 3000 ≈ 2995. The answer checks.

<u>State</u>. A total of 2995 cubic centimeters of liquid was poured.

7. <u>Familiarize</u>. We visualize the situation. We let x = the amount by which the water loss from the skin exceeds the water loss from the lungs.

Water loss from lungs 400 cubic centimeters	Excess loss from skin x
Water loss from skin 500 cubic centimeters	

<u>Translate</u>. This is a "how-much-more" situation. Translate to an equation.

Water lost from lungs	plus	Additional amount of water	is	Water lost from skin
400	+	x	=	500

<u>Solve</u>. Solve the equation.

$$400 + x = 500$$
$$400 + x - 400 = 500 - 400 \quad \text{Subtracting 400 on both sides}$$
$$x = 100$$

<u>Check</u>. We can add the result to 400: 400 + 100 = 500. The answer checks.

<u>State</u>. 100 cubic centimeters more water is lost from the skin than from the lungs.

9. <u>Familiarize</u>. We visualize the situation. We let p = the number by which the population of the Tokyo-Yokohama area exceeds the population of the New York-northeastern New Jersey area.

NY – NJ population 17,013,000	Excess p
Tokyo-Yokohama population 17,317,000	

<u>Translate</u>. This is a "how-much-more" situation. Translate to an equation.

People in New York - northeastern New Jersey area	plus	How many more people	is	People in Tokyo - Yokohama area
17,013,000	+	p	=	17,317,000

<u>Solve</u>. We solve the equation.

$$17,013,000 + p = 17,317,000$$
$$17,013,000 + p - 17,013,000 = 17,317,000 - 17,013,000$$
$$\text{Subtracting 17,013,000 on both sides}$$
$$p = 304,000$$

<u>Check</u>. We can add the result to 17,013,000: 17,013,000 + 304,000 = 17,317,000. We can also estimate: 17,317,000 - 17,013,000 ≈ 17,300,000 - 17,000,000 = 300,000 ≈ 304,000. The answer checks.

<u>State</u>. There are 304,000 more people in the Tokyo-Yokohama area.

11. <u>Familiarize</u>. This is a multistep problem.

To find the new balance, find the total amount of the three checks. Then take that amount away from the original balance. We visualize the situation. We let t = the total amount of the three checks and n = the new balance.

$246			
$45	$78	$32	n
t			n

<u>Translate</u>. To find the total amount of the three checks, write an addition sentence.

Amount of first check	plus	Amount of second check	plus	Amount of third check	is	Total Amount
$45	+	$78	+	$32	=	t

11. (continued)

Solve. To solve the equation we carry out the addition.

$$\begin{array}{r} \$\ 4\ 5 \\ 7\ 8 \\ +\ \ 3\ 2 \\ \hline \$1\ 5\ 5 \end{array}$$

The total amount of the three checks is $155.

To find the new balance, we have a "take-away" situation.

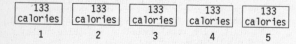

Original balance	minus	Amount of checks	is	New balance
↓	↓	↓	↓	↓
$246	−	$155	=	n

To solve we carry out the subtraction.

$$\begin{array}{r} \$\ 2\ 4\ 6 \\ -\ 1\ 5\ 5 \\ \hline \$\ \ \ 9\ 1 \end{array}$$

Check. We add the amounts of the three checks and the new balance. The sum should be the original balance.

$45 + $78 + $32 + $91 = $246

The answer checks.

State. The new balance is $91.

13. Familiarize. We first make a drawing. Repeated addition works well here. We let n = the number of calories burned in 5 hours.

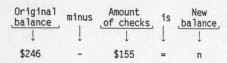

Translate. Translate to a number sentence.

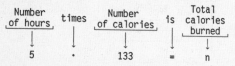

Number of hours	times	Number of calories	is	Total calories burned
↓	↓	↓	↓	↓
5	·	133	=	n

Solve. To solve the equation, we carry out the multiplication.

5 · 133 = n

665 = n   Doing the multiplication

Check. We can repeat the calculation. We can also estimate: 5·133 ≈ 5·130 = 650 ≈ 665. The answer checks.

State. In 5 hours, 665 calories would be burned.

15. Familiarize. We first make a drawing. Repeated addition works well here. We let x = the number of seconds in an hour.

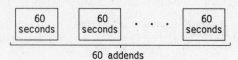

60 addends

Translate. We translate to an equation.

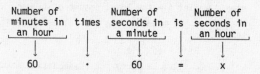

Number of minutes in an hour	times	Number of seconds in a minute	is	Number of seconds in an hour
↓	↓	↓	↓	↓
60	·	60	=	x

Solve. We carry out the multiplication.

60·60 = x

3600 = x   Doing the multiplication

Check. We can check by repeating the calculation. The answer checks.

State. There are 3600 seconds in an hour.

17. We first make a drawing.

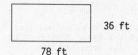

78 ft

Translate. Using the formula for area, we have A = ℓ·w = 78·36.

Solve. Carry out the multiplication.

$$\begin{array}{r} 7\ 8 \\ \times\ \ 3\ 6 \\ \hline 4\ 6\ 8 \\ 2\ 3\ 4\ 0 \\ \hline 2\ 8\ 0\ 8 \end{array}$$

Thus, A = 2808.

Check. We repeat the calculation. The answer checks.

State. The area is 2808 sq ft.

19. Familiarize. We visualize the situation. We let d = the diameter of the earth.

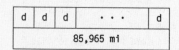

85,965 mi

Translate. Repeated addition applies here. The following multiplication corresponds to the situation.

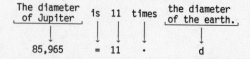

The diameter of Jupiter	is	11	times	the diameter of the earth.
↓	↓	↓	↓	↓
85,965	=	11	·	d

<u>19.</u> (continued)

   <u>Solve.</u> To solve the equation, we divide by 11 on both sides.

$$85,965 = 11 \cdot d$$

$$\frac{85,965}{11} = \frac{11 \cdot d}{11}$$

$$7815 = d$$

<u>Check.</u> To check we multiply 7815 by 11:

$$11 \cdot 7815 = 85,965$$

This checks.

<u>State.</u> The diameter of the earth is 7815 mi.

<u>21.</u> <u>Familiarize.</u> This is a multistep problem.

We must find the total cost of the 8 suits and the total cost of the 3 shirts. The amount spent is the sum of these two totals.

Repeated addition works well in finding the total cost of the 8 suits and the total cost of the 3 shirts. We let x = the total cost of the suits and y = the total cost of the shirts.

| $195 | $195 | • • • | $195 |

8 addends

| $26 | $26 | $26 |

3 addends

<u>Translate.</u> We translate to two equations.

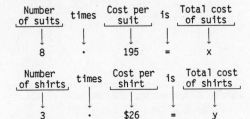

| Number of suits | times | Cost per suit | is | Total cost of suits |
| 8 | • | 195 | = | x |

| Number of shirts | times | Cost per shirt | is | Total cost of shirts |
| 3 | • | $26 | = | y |

<u>Solve.</u> To solve the equations, we carry out the multiplications.

```
    1 9 5
  ×     8
  1 5 6 0     Thus, x = $1560.
```

```
      2 6
    ×  3
    7 8     Thus, y = $78.
```

We let a = the total amount spent.

| Total cost of suits | plus | Total cost of shirts | is | Amount spent |
| $1560 | + | $78 | = | a |

To solve the equation carry out the addition.

```
  $ 1 5 6 0
  +     7 8
  $ 1 6 3 8
```

<u>21.</u> (continued)

   <u>Check.</u> We repeat the calculations. The answer checks.

<u>State.</u> The amount spent is $1638.

<u>23.</u> <u>Familiarize.</u> This is a multistep problem.

We must find the area of the lot and the area of the garden. Then we take the area of the garden away from the area of the lot. We let A = the area of the lot and G = the area of the garden.

First, make a drawing of the lot with the garden. The area left over is shaded.

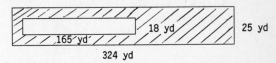

324 yd

<u>Translate.</u> We use the formula for area twice.

$$A = \ell \cdot w = 324 \cdot 25$$

$$G = \ell \cdot w = 185 \cdot 18$$

<u>Solve.</u> We carry out the multiplications.

```
    3 2 4
  ×   2 5
  1 6 2 0
  6 4 8 0
  8 1 0 0
```

A = 8100 sq yd.

```
    1 6 5
  ×   1 8
  1 3 2 0
  1 6 5 0
  2 9 7 0
```

G = 2970 sq yd.

To find the area left over we have a "take-away" situation. We let a = the area left over.

| Area of lot | minus | Area of garden | is | Area left over |
| 8100 | - | 2970 | = | a |

To solve, carry out the subtraction.

```
    8 1 0 0
  - 2 9 7 0
    5 1 3 0
```

<u>Check.</u> We repeat the calculations. The answer checks.

<u>State.</u> The area left over is 5130 sq yd.

25. <u>Familiarize</u>. We first draw a picture. We let n = the number of bottles to be filled.

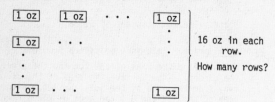

<u>Translate and Solve</u>. We translate to an equation and solve as follows:

$$608 \div 16 = n$$

$$
\begin{array}{r}
3\,8 \\
16\,\overline{\smash{)}\,608} \\
\underline{4\,8\,0} \\
1\,2\,8 \\
\underline{1\,2\,8} \\
0
\end{array}
$$

<u>Check</u>. We can check by multiplying the number of bottles by 16: 16·38 = 608. The answer checks.

<u>State</u>. 38 sixteen-oz bottles can be filled.

27. <u>Familiarize</u>. We draw a picture. We let y = the number of rows.

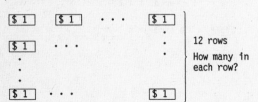

15 in each row.

How many rows?

<u>Translate and Solve</u>. We translate to an equation and solve as follows:

$$225 \div 15 = y$$

$$
\begin{array}{r}
1\,5 \\
15\,\overline{\smash{)}\,225} \\
\underline{1\,5\,0} \\
7\,5 \\
\underline{7\,5} \\
0
\end{array}
$$

<u>Check</u>. We can check by multiplying the number of rows by 15: 15·15 = 225. The answer checks.

<u>State</u>. There are 15 rows.

29. <u>Familiarize</u>. We first draw a picture. We let x = the amount of each payment.

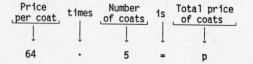

12 rows

How many in each row?

29. (continued)

<u>Translate and Solve</u>. We translate to an equation and solve as follows:

$$324 \div 12 = x$$

$$
\begin{array}{r}
2\,7 \\
12\,\overline{\smash{)}\,324} \\
\underline{2\,4\,0} \\
8\,4 \\
\underline{8\,4} \\
0
\end{array}
$$

<u>Check</u>. We can check by multiplying 27 by 12: 12·27 = 324. the answer checks.

<u>State</u>. Each payment is $27.

31. <u>Familiarize</u>. We draw a picture. We let n = the number of bags to be filled.

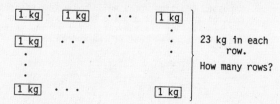

23 kg in each row.

How many rows?

<u>Translate and Solve</u>. We translate to an equation and solve as follows:

$$885 \div 23 = n$$

$$
\begin{array}{r}
3\,8 \\
23\,\overline{\smash{)}\,885} \\
\underline{6\,9\,0} \\
1\,9\,5 \\
\underline{1\,8\,4} \\
1\,1
\end{array}
$$

<u>Check</u>. We can check by multiplying the number of bags by 23 and adding the remainder of 11:

$$23 \cdot 38 = 874$$

$$874 + 11 = 885$$

The answer checks.

<u>State</u>. 38 twenty-three-kg bags can be filled. There will be 11 kg of sand left over.

33. <u>Familiarize</u>. This is a multistep problem.

We must find the total price of the 5 coats. Then we must find how many 20's there are in the total price. Let p = the total price of the coats.

To find the total price of the 5 coats we can use repeated addition.

| $64 | $64 | $64 | $64 | $64 |

5 addends

<u>Translate</u>.

Price per coat	times	Number of coats	is	Total price of coats
↓	↓	↓	↓	↓
64	·	5	=	p

33. (continued)

Solve. First we carry out the multiplication.

$64 \cdot 5 = p$

$320 = p$

The total price of the 5 coats is $320. Repeated addition can be used again to find how many 20's in $320. We let x = the number of $20 bills required.

$320			
$20	$20	$\cdots$	$20

Translate to an equation and solve.

$20 \cdot x = 320$

$\dfrac{20 \cdot x}{20} = \dfrac{320}{20}$ Dividing on both sides by 20

$x = 16$

Check. We repeat the calculations. The answer checks.

Solve. It took 16 twenty dollar bills.

35. Familiarize. To find how far apart on the map the cities are we must find how many 55's there are in 605. Repeated addition applies here. We let d = the distance between the cities on the map.

605 mi			
55 mi	55 mi	$\cdots$	55 mi

Translate.

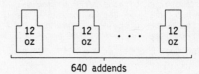

$$55 \cdot d = 605$$

Solve. We divide on both sides by 55.

$55 \cdot d = 605$

$\dfrac{55 \cdot d}{55} = \dfrac{605}{55}$ Dividing on both sides by 55

$d = 11$

Check. We multiply the number of inches by 55: $11 \cdot 55 = 605$. The answer checks.

State. The cities are 11 inches apart on the map.

Familiarize. When two cities are 14 in. apart on the map, we can use repeated addition to find how far apart they are in reality. We let n = the distance between the cities in reality.

Since 1 in. represents 55 mi, we can draw the following picture:

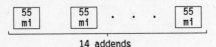

14 addends

35. (continued)

Translate.

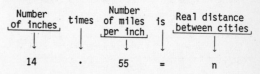

Solve. We carry out the multiplication.

$14 \times 55 = n$

$770 = n$ Doing the multiplication

Check. We repeat the calculation. The answer checks.

State. The cities are 770 mi apart in reality.

37. Familiarize. This is a multistep problem.

We must first find the total number of ounces of soda. Then we must find how many 16's there are in that total. We let x = the total number of ounces of soda.

To find the total number of ounces of soda we can use repeated addition.

12 oz	12 oz	$\cdots$	12 oz

640 addends

Translate.

Solve. We carry out the multiplication.

$640 \times 12 = x$

$7680 = x$ Doing the multiplication

There are 7680 oz of soda to be bottled. We can also use repeated addition to find how many 16's there are in 7680. Let y = this number.

7680 oz			
16 oz	16 oz	$\cdots$	16 oz

Translate to an equation and solve.

$7680 \div 16 = y$

$480 = y$ Doing the division

Check. If we multiply 480 by 16, we get 7680, and $7680 \div 12 = 640$. The answer checks.

State. The soda will fill 480 sixteen-oz bottles.

39. <u>Familiarize</u>. This is a multistep problem.

We must find how many 100's there are in 3500. Then we must find that number times 15.

We first draw a picture.

One pound			
3500 calories			
100 cal	100 cal	· · ·	100 cal
15 min	15 min	· · ·	15 min

In Example 11 it was determined that there are 35 100's in 3500. We let t = the time you have to bicycle to lose a pound.

<u>Translate and Solve</u>. We know that bicycling at 9 mph for 15 min burns off 100 calories, so we need to bicycle for 35 times 15 min in order to burn off one pound. Translate to an equation and solve. We let t = the time required to lose one pound by bicycling.

$$35 \times 15 = t$$
$$525 = t$$

<u>Check</u>. Suppose you bicycle for 525 minutes. If we divide 525 by 15, we get 35, and 35 times 100 is 3500, the number of calories that must be burned off to lose one pound. The answer checks.

<u>State</u>. You must bicycle at 9 mph for 525 min, or 8 hr 45 min, to lose one pound.

41. <u>Familiarize</u>. We visualize the situation. Let d = the distance light travels in 1 sec.

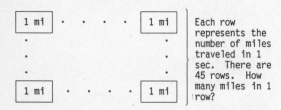

Each row represents the number of miles traveled in 1 sec. There are 45 rows. How many miles in 1 row?

<u>Translate and Solve</u>. We translate to an equation and solve as follows:

$$8,370,000 \div 45 = d$$

```
            1 8 6,0 0 0
      4 5 ) 8,3 7 0,0 0 0
            4,5 0 0,0 0 0
            3,8 7 0,0 0 0
            3,6 0 0,0 0 0
              2 7 0,0 0 0
              2 7 0,0 0 0
                        0
```

<u>Check</u>. We can check by multiplying the number of miles traveled in 1 sec by 45 sec:

$$45 \cdot 186,000 = 8,370,000$$

The answer checks.

<u>State</u>. Light travels 186,000 mi in 1 sec.

Exercise 1.9

1. Exponential notation for $3 \cdot 3 \cdot 3 \cdot 3$ is 3^4.

3. Exponential notation for $5 \cdot 5$ is 5^2.

5. Exponential notation for $7 \cdot 7 \cdot 7 \cdot 7 \cdot 7$ is 7^5.

7. Exponential notation for $10 \cdot 10 \cdot 10$ is 10^3.

9. $7^2 = 7 \cdot 7 = 49$

11. $9^3 = 9 \cdot 9 \cdot 9 = 729$

13. $12^4 = 12 \cdot 12 \cdot 12 \cdot 12 = 20,736$

15. $11^2 = 11 \cdot 11 = 121$

17. $12 + (6 + 4) = 12 + 10$ Doing the calculation inside parentheses

$= 22$ Adding

19. $52 - (40 - 8) = 52 - 32$ Doing the calculation inside parentheses

$= 20$ Subtracting

21. $1000 \div (100 \div 10) = 1000 \div 10$ Doing the calculation inside parentheses

$= 100$ Dividing

23. $(256 \div 64) \div 4 = 4 \div 4$ Doing the calculation inside parentheses

$= 1$ Dividing

25. $(2 + 5)^2 = 7^2$ Doing the calculation inside parentheses

$= 49$ Evaluating the exponential expression

27. $2 + 5^2 = 2 + 25$ Evaluating the exponential expression

$= 27$ Adding

29. $16 \cdot 24 + 50 = 384 + 50$ Doing all multiplications and divisions in order from left to right

$= 434$ Doing all additions and subtractions in order from left to right

31. $83 - 7 \cdot 6 = 83 - 42$ Doing all multiplications and divisions in order from left to right

$= 41$ Doing all additions and subtractions in order from left to right

33. $10 \cdot 10 - 3 \times 4 = 100 - 12$ Doing all multiplications and divisions in order from left to right

           $= 88$ Doing all additions and subtractions in order from left to right

35. $4^3 \div 8 - 4 = 64 \div 8 - 4$ Evaluating the exponential expression

           $= 8 - 4$ Doing all multiplications and divisions in order from left to right

           $= 4$ Doing all additions and subtractions in order from left to right

37. $17 \cdot 20 - (17 + 20) = 17 \cdot 20 - 37$ Carrying out operations inside parentheses

           $= 340 - 37$ Doing all multiplications and divisions in order from left to right

           $= 303$ Doing all additions and subtractions in order from left to right

39. $6 \cdot 10 - 4 \cdot 10 = 60 - 40$ Doing all multiplications and divisions in order from left to right

           $= 20$ Doing all additions and subtractions in order from left to right

41. $300 \div 5 + 10 = 60 + 10$ Doing all multiplications and divisions in order from left to right

           $= 70$ Doing all additions and subtractions in order from left to right

43. $3 \cdot (2 + 8)^3 - 5 \cdot (4 - 3)^2 = 3 \cdot 10^2 - 5 \cdot 1^2$

           Carrying out operations inside parentheses

           $= 3 \cdot 100 - 5 \cdot 1$

           Evaluating the exponential expressions

           $= 300 - 5$

           Doing all multiplications and divisions in order from left to right

           $= 295$

           Doing all additions and subtractions in order from left to right

45. $4^2 + 8^2 \div 2^2 = 16 + 64 \div 4$

           $= 16 + 16$

           $= 32$

47. $10^3 - 10 \cdot 6 - (4 + 5 \cdot 6)$

      $= 10^3 - 10 \cdot 6 - (4 + 30)$

      $= 10^3 - 10 \cdot 6 - 34$

      $= 1000 - 10 \cdot 6 - 34$

      $= 1000 - 60 - 34$

      $= 906$

49. $6 \times 11 - (7 + 3) \div 5 - (6 - 4)$

      $= 6 \times 11 - 10 \div 5 - 2$

      $= 66 - 2 - 2$

      $= 64 - 2$

      $= 62$

51. $120 - 3^3 \cdot 4 \div (30 - 24)$

      $= 120 - 3^3 \cdot 4 \div 6$

      $= 120 - 27 \cdot 4 \div 6$

      $= 120 - 108 \div 6$

      $= 120 - 18$

      $= 102$

53. $8 \times 13 + \{42 \div [18 - (6 + 5)]\}$

      $= 8 \times 13 + \{42 \div [18 - 11]\}$

      $= 8 \times 13 + \{42 \div 7\}$

      $= 8 \times 13 + 6$

      $= 104 + 6$

      $= 110$

55. $[14 - (3 + 5) \div 2] - [18 \div (8 - 2)]$

      $= [14 - 8 \div 2] - [18 \div 6]$

      $= [14 - 4] - 3$

      $= 10 - 3$

      $= 7$

57. $(82 - 14) \times [(10 + 45 \div 5) - (6 \cdot 6 - 5 \cdot 5)]$

      $= (82 - 14) \times [(10 + 9) - (36 - 25)]$

      $= (82 - 14) \times [19 - 11]$

      $= 68 \times 8$

      $= 544$

59. $4 \times \{(200 - 50 \div 5) - [(35 \div 7) \cdot (35 \div 7) - 4 \times 3]\}$

      $= 4 \times \{(200 - 10) - [5 \cdot 5 - 4 \times 3]\}$

      $= 4 \times \{190 - [5 \cdot 5 - 4 \times 3]\}$

      $= 4 \times \{190 - [25 - 12]\}$

      $= 4 \times \{190 - 13\}$

      $= 4 \times 177$

      $= 708$

<u>61.</u> $1 + 5 \cdot 4 + 3 = 1 + 20 + 3$

                 $= 24$ Correct answer

To make the incorrect answer correct we add parentheses:

    $1 + 5 \cdot (4 + 3) = 36$

<u>63.</u> $12 \div 4 + 2 \cdot 3 - 2 = 3 + 6 - 2$

                     $= 7$ Correct answer

To make the incorrect answer correct we add parentheses:

    $12 \div (4 + 2) \cdot 3 - 2 = 4$

Exercise Set 2.1

1. We first find some factorizations:

    16 = 1·16   16 = 8·2

    16 = 2·8   16 = 16·1

    16 = 4·4

    Factors: 1, 2, 4, 8, 16

3. We first find some factorizations:

    54 = 1·54   54 = 9·6

    54 = 2·27   54 = 18·3

    54 = 3·18   54 = 27·2

    54 = 6·9   54 = 54·1

    Factors: 1, 2, 3, 6, 9, 18, 27, 54

5. We first find some factorizations:

    4 = 1·4   4 = 2·2

    Factors: 1, 2, 4

7. The only factorization is 7 = 1·7.

    Factors: 1, 7

9. The only factorization is 1·1.

    Factor: 1

11. We first find some factorizations:

    98 = 1·98   98 = 7·14

    98 = 2·49

    Factors: 1, 2, 7, 14, 49, 98

13.

1·4 = 4	6·4 = 24
2·4 = 8	7·4 = 28
3·4 = 12	8·4 = 32
4·4 = 16	9·4 = 36
5·4 = 20	10·4 = 40

15.

1·20 = 20	6·20 = 120
2·20 = 40	7·20 = 140
3·20 = 60	8·20 = 160
4·20 = 80	9·20 = 180
5·20 = 100	10·20 = 200

17.

1·3 = 3	6·3 = 18
2·3 = 6	7·3 = 21
3·3 = 9	8·3 = 24
4·3 = 12	9·3 = 27
5·3 = 15	10·3 = 30

19.

1·12 = 12	6·12 = 72
2·12 = 24	7·12 = 84
3·12 = 36	8·12 = 96
4·12 = 48	9·12 = 108
5·12 = 60	10·12 = 120

21.

1·10 = 10	6·10 = 60
2·10 = 20	7·10 = 70
3·10 = 30	8·10 = 80
4·10 = 40	9·10 = 90
5·10 = 50	10·10 = 100

23.

1·9 = 9	6·9 = 54
2·9 = 18	7·9 = 63
3·9 = 27	8·9 = 72
4·9 = 36	9·9 = 81
5·9 = 45	10·9 = 90

25. We divide 26 by 6.

```
      4
  6 ) 2 6
      2 4
        2
```

Since the remainder is not 0, 26 is not divisible by 6.

27. We divide 1880 by 8.

```
        2 3 5
  8 ) 1 8 8 0
      1 6 0 0
        2 8 0
        2 4 0
          4 0
          4 0
            0
```

Since the remainder is 0, 1880 is divisible by 8.

29. We divide 256 by 16.

```
           1 6
  1 6 ) 2 5 6
        1 6 0
          9 6
          9 6
            0
```

Since the remainder is 0, 256 is divisible by 16.

31. We divide 4227 by 9.

```
          4 6 9
  9 ) 4 2 2 7
      3 6 0 0
        6 2 7
        5 4 0
          8 7
          8 1
            6
```

Since the remainder is not 0, 4227 is not divisible by 9.

33. We divide 8650 by 16.

$$
\begin{array}{r}
540 \\
16\overline{)8650} \\
\underline{8000} \\
650 \\
\underline{640} \\
10
\end{array}
$$

Since the remainder is not 0, 8650 is not divisible by 16.

35. 1 is <u>neither</u> prime nor composite.

37. The number 9 has factors 1, 3, and 9.

Since 9 is not 1 and not prime, it is <u>composite</u>.

39. The number 11 is <u>prime</u>. It has only the factors 1 and 11.

41. The number 29 is <u>prime</u>. It has only the factors 1 and 29.

43.
$$
\begin{array}{r}
2 \\
2\overline{)4} \\
2\overline{)8}
\end{array}
$$
← 2 is prime

$8 = 2 \cdot 2 \cdot 2$

45.
$$
\begin{array}{r}
7 \\
2\overline{)14}
\end{array}
$$
← 7 is prime

$14 = 2 \cdot 7$

47.
$$
\begin{array}{r}
11 \\
2\overline{)22}
\end{array}
$$
← 11 is prime

$22 = 2 \cdot 11$

49.
$$
\begin{array}{r}
5 \\
5\overline{)25}
\end{array}
$$
← 5 is prime
(25 is not divisible by 2 or 3. We move to 5.)

$25 = 5 \cdot 5$

51.
$$
\begin{array}{r}
5 \\
5\overline{)25} \\
2\overline{)50}
\end{array}
$$
← 5 is prime
(25 is not divisible by 2 or 3. We move to 5.)

$50 = 2 \cdot 5 \cdot 5$

53.
$$
\begin{array}{r}
13 \\
13\overline{)169} \\
\underline{169} \\
0
\end{array}
$$
← 13 is prime
(169 is not divisible by 2, 3, 5, 7 and 11. We move to 13.)

$169 = 13 \cdot 13$

55.
$$
\begin{array}{r}
5 \\
5\overline{)25} \\
2\overline{)50} \\
2\overline{)100}
\end{array}
$$
← 5 is prime
(25 is not divisible by 2 or 3. We move to 5.)

$100 = 2 \cdot 2 \cdot 5 \cdot 5$

We can also use a factor tree.

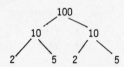

57.
$$
\begin{array}{r}
7 \\
5\overline{)35}
\end{array}
$$
← 7 is prime
(35 is not divisible by 2 or 3. We move to 5.)

$35 = 5 \cdot 7$

59.
$$
\begin{array}{r}
3 \\
3\overline{)9} \\
2\overline{)18} \\
2\overline{)36} \\
2\overline{)72}
\end{array}
$$
← 3 is prime
(9 is not divisible by 2. We move to 3.)

$72 = 2 \cdot 2 \cdot 2 \cdot 3 \cdot 3$

We can also use a factor tree, as shown in Example 9 in the text.

61.
$$
\begin{array}{r}
11 \\
7\overline{)77}
\end{array}
$$
← 11 is prime
(77 is not divisible by 2, 3, or 5. We move to 7.)

$77 = 7 \cdot 11$

63.
$$
\begin{array}{r}
7 \\
2\overline{)14} \\
2\overline{)28} \\
2\overline{)56} \\
2\overline{)112}
\end{array}
$$
← 7 is prime

$112 = 2 \cdot 2 \cdot 2 \cdot 2 \cdot 7$

We can also use a factor tree.

65.
$$
\begin{array}{r}
5 \\
5\overline{)25} \\
3\overline{)75} \\
2\overline{)150} \\
2\overline{)300}
\end{array}
$$
← 5 is prime

$300 = 2 \cdot 2 \cdot 3 \cdot 5 \cdot 5$

65. (continued)

We can also use a factor tree.

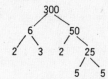

67.
```
  1 3
×   2
  2 6
```
2·13 = 26

69.
```
      0
2 2⌐0
      0
      0
0 ÷ 22 = 0
```

71. Since there are 2 factors a rectangular array is indicated. One factor gives the number of rows, the other the number of objects in each row. Thus, we have a rectangular array of 6 rows with 9 objects each or 9 rows with 6 objects each.

Exercise Set 2.2

---

1. A number is divisible by 2 if its <u>ones digit</u> is even.

    4<u>6</u> is divisible by 2 because <u>6</u> is even.

    22<u>4</u> is divisible by 2 because <u>4</u> is even.

    1<u>9</u> is not divisible by 2 because <u>9</u> is not even.

    55<u>5</u> is not divisible by 2 because <u>5</u> is not even.

    30<u>0</u> is divisible by 2 because <u>0</u> is even.

    3<u>6</u> is divisible by 2 because <u>6</u> is even.

    45,27<u>0</u> is divisible by 2 because <u>0</u> is even.

    444<u>4</u> is divisible by 2 because <u>4</u> is even.

    8<u>5</u> is not divisible by 2 because <u>5</u> is not even.

    71<u>1</u> is not divisible by 2 because <u>1</u> is not even.

    13,25<u>1</u> is not divisible by 2 because <u>1</u> is not even.

    254,76<u>5</u> is not divisible by 2 because <u>5</u> is not even.

3. A number is divisible by 4 if the <u>number</u> named by the last <u>two</u> digits is divisible by 4.

    <u>46</u> is not divisible by 4 because <u>46</u> is not divisible by 4.

    2<u>24</u> is divisible by 4 because <u>24</u> is divisible by 4.

    <u>19</u> is not divisible by 4 because <u>19</u> is not divisible by 4.

    5<u>55</u> is not divisible by 4 because <u>55</u> is not divisible by 4.

    3<u>00</u> is divisible by 4 because <u>00</u> is divisible by 4.

    <u>36</u> is divisible by 4 because <u>36</u> is divisible by 4.

    45,2<u>70</u> is not divisible by 4 because <u>70</u> is not divisible by 4.

    44<u>44</u> is divisible by 4 because <u>44</u> is divisible by 4.

    <u>85</u> is not divisible by 4 because <u>85</u> is not divisible by 4.

    7<u>11</u> is not divisible by 4 because <u>11</u> is not divisible by 4.

    13,2<u>51</u> is not divisible by 4 because <u>51</u> is not divisible by 4.

    254,7<u>65</u> is not divisible by 4 because <u>65</u> is not divisible by 4.

5. For a number to be divisible by 6, the sum of the digits must be divisible by 3 and the ones digit must be 0, 2, 4, 6 or 8 (even). It is most efficient to determine if the ones digit is even first and then, if so, to determine if the sum of the digits is divisible by 3.

    46 is not divisible by 6 because 46 is not divisible by 3.

    4 + 6 = 10
          ↑
       Not divisible by 3

    224 is not divisible by 6 because 224 is not divisible by 3.

    2 + 2 + 4 = 8
              ↑
          Not divisible by 3

    19 is not divisible by 6 because 19 is not divisible by 2.

    19
    ↑
    Not even

    555 is not divisible by 6 because 555 is not divisible by 2.

    555
    ↑
    Not even

5. (continued)

300 is divisible by 6.

300   3 + 0 + 0 = 3
↑   ↑
Even   Divisible by 3

36 is divisible by 6.

36   3 + 6 = 9
↑   ↑
Even   Divisible by 3

45,270 is divisible by 6.

45,270   4 + 5 + 2 + 7 + 0 = 18
↑
Even   Divisible by 3

4444 is not divisible by 6 because 444 is not divisible by 3.

4 + 4 + 4 + 4 = 16

Not divisible by 3

85 is not divisible by 6 because 85 is not divisible by 2.

85
↑
Not even

711 is not divisible by 6 because 711 is not divisible by 2.

711
↑
Not even

13,251 is not divisible by 6 because 13,251 is not divisible by 2.

13,251
↑
Not even

254,765 is not divisible by 6 because 254,765 is not divisible by 2.

254,765
↑
Not even

7. A number is divisible by 9 if the sum of the digits is divisible by 9.

46 is not divisible by 9 because 4 + 6 = 10 and 10 is not divisible by 9.

224 is not divisible by 9 because 2 + 2 + 4 = 8 and 8 is not divisible by 9.

19 is not divisible by 9 because 1 + 9 = 10 and 10 is not divisible by 9.

555 is not divisible by 9 because 5 + 5 + 5 = 15 and 15 is not divisible by 9.

300 is not divisible by 9 because 3 + 0 + 0 = 3 and 3 is not divisible by 9.

36 is divisible by 9 because 3 + 6 = 9 and 9 is divisible by 9.

45,270 is divisible by 9 because 4 + 5 + 2 + 7 + 0 = 18 and 18 is divisible by 9.

7. (continued)

4444 is not divisible by 9 because 4 + 4 + 4 + 4 = 16 and 16 is not divisible by 9.

85 is not divisible by 9 because 8 + 5 = 13 and 13 is not divisible by 9.

711 is divisible by 9 because 7 + 1 + 1 = 9 and 9 is divisible by 9.

13,251 is not divisible by 9 because 1 + 3 + 2 + 5 + 1 = 12 and 12 is not divisible by 9.

254,765 is not divisible by 9 because 2 + 5 + 4 + 7 + 6 + 5 = 29 and 29 is not divisible by 9.

9. A number is divisible by 3 if the sum of the digits is divisible by 3.

56 is not divisible by 3 because 5 + 6 = 11 and 11 is not divisible by 3.

324 is divisible by 3 because 3 + 2 + 4 = 9 and 9 is divisible by 3.

784 is not divisible by 3 because 7 + 8 + 4 = 19 and 19 is not divisible by 3.

55,555 is not divisible by 3 because 5 + 5 + 5 + 5 + 5 = 25 and 25 is not divisible by 3.

200 is not divisible by 3 because 2 + 0 + 0 = 2 and 2 is not divisible by 3.

42 is divisible by 3 because 4 + 2 = 6 and 6 is divisible by 3.

501 is divisible by 3 because 5 + 0 + 1 = 6 and 6 is divisible by 3.

3009 is divisible by 3 because 3 + 0 + 0 + 9 = 12 and 12 is divisible by 3.

75 is divisible by 3 because 7 + 5 = 12 and 12 is divisible by 3.

812 is not divisible by 3 because 8 + 1 + 2 = 11 and 11 is not divisible by 3.

2345 is not divisible by 3 because 2 + 3 + 4 + 5 = 14 and 14 is not divisible by 3.

2001 is divisible by 3 because 2 + 0 + 0 + 1 = 3 and 3 is divisible by 3.

11. A number is divisible by 5 if the ones digit is 0 or 5.

56 is not divisible by 5 because the ones digit (6) is not 0 or 5.

324 is not divisible by 5 because the ones digit (4) is not 0 or 5.

784 is not divisible by 5 because the ones digit (4) is not 0 or 5.

55,555 is divisible by 5 because the ones digit (5) is 5.

200 is divisible by 5 because the ones digit (0) is 0.

11. (continued)

    4<u>2</u> is not divisible by 5 because the ones digit
    (2) is not 0 or 5.

    50<u>1</u> is not divisible by 5 because the ones digit
    (1) is not 0 or 5.

    300<u>9</u> is not divisible by 5 because the ones digit
    (9) is not 0 or 5.

    7<u>5</u> is divisible by 5 because the ones digit (5)
    is 5.

    81<u>2</u> is not divisible by 5 because the ones digit
    (2) is not 0 or 5.

    234<u>5</u> is divisible by 5 because the ones digit (5)
    is 5.

    200<u>1</u> is not divisible by 5 because the ones digit
    (1) is not 0 or 5.

13. A number is divisible by 9 if the sum of the
    digits is divisible by 9.

    56 is not divisible by 9 because 5 + 6 = 11 and 11
    is not divisible by 9.

    324 is divisible by 9 because 3 + 2 + 4 = 9 and 9
    is divisible by 9.

    784 is not divisible by 9 because 7 + 8 + 4 = 19
    and 19 is not divisible by 9.

    55,555 is not divisible by 9 because 5 + 5 + 5 +
    5 + 5 = 25 and 25 is not divisible by 9.

    200 is not divisible by 9 because 2 + 0 + 0 = 2
    and 2 is not divisible by 9.

    42 is not divisible by 9 because 4 + 2 = 6 and 6
    is not divisible by 9.

    501 is not divisible by 9 because 5 + 0 + 1 = 6
    and 6 is not divisible by 9.

    3009 is not divisible by 9 because 3 + 0 + 0 +
    9 = 12 and 12 is not divisible by 9.

    75 is not divisible by 9 because 7 + 5 = 12 and 12
    is not divisible by 9.

    812 is not divisible by 9 because 8 + 1 + 2 = 11
    and 11 is not divisible by 9.

    2345 is not divisible by 9 because 2 + 3 + 4 + 5 =
    14 and 14 is not divisible by 9.

    2001 is not divisible by 9 because 2 + 0 + 0 +
    1 = 3 and 3 is not divisible by 9.

15. A number is divisible by 10 if the ones digit is
    0.

    Of the numbers under consideration, the only one
    whose ones digit is 0 is 200. Therefore, 200 is
    divisible by 10. None of the other numbers is
    divisible by 10.

17. $56 + x = 194$

    $56 + x - 56 = 194 - 56$ Subtracting 56 on both
                                 sides

              $x = 138$

    The solution is 138.

19. <u>Familiarize</u>. This is a multistep problem. Find
    the total cost of the shirts and the total cost of
    the trousers and then find the sum of the two.

    We let s = the total cost of the shirts and
    t = the total cost of the trousers.

    <u>Translate</u>. We write two equations.

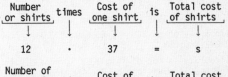

Number of shirts	times	Cost of one shirt	is	Total cost of shirts
↓	↓	↓	↓	↓
12	·	37	=	s

Number of pairs of trousers	times	Cost of one pair	is	Total cost of trousers
↓	↓	↓	↓	↓
4	·	59	=	t

    <u>Solve</u>. We carry out the multiplication.

        $12·37 = s$

           $444 = s$ Doing the multiplication

    The total cost of the 12 shirts is $444.

         $4·59 = t$

           $236 = t$ Doing the multiplication

    The total cost of the 4 pairs of trousers is $236.

    Now we find the total amount spent.

Total cost of shirts	plus	Total cost of trousers	is	Total amount spent
↓	↓	↓	↓	↓
444	+	236	=	a

    To solve the equation, carry out the addition.

              $4\ 4\ 4$
          $+\ 2\ 3\ 6$
          _____
              $6\ 8\ 0$

    <u>Check</u>. We can repeat the calculations. The
    answer checks.

    <u>State</u>. The total cost is $680.

21. 7800 is divisible by 2 because the ones digit (0)
    is even.

    $7800 ÷ 2 = 3900$ so $7800 = 2·3900$.

    3900 is divisible by 2 because the ones digit (0)
    is even.

    $3900 ÷ 2 = 1950$ so $3900 = 2·1950$ and $7800 = 2·2·1950$.

    1950 is divisible by 2 because the ones digit (0)
    is even.

    $1950 ÷ 2 = 975$ so $1950 = 2·975$ and $7800 = 2·2·2·975$.

21. (continued)

975 is not divisible by 2 because the ones digit (5) is not even. Move on to 3.

975 is divisible by 3 because the sum of the digits (9 + 7 + 5 = 21) is divisible by 3.

975 ÷ 3 = 325 so 975 = 3·325 and 7800 = 2·2·2·3·325.

Since 975 is not divisible by 2, none of its factors is divisible by 2. Therefore, we no longer need to check for divisibility by 2.

325 is not divisible by 3 because the sum of the digits (3 + 2 + 5 = 10) is not divisible by 3. Move on to 5.

325 is divisible by 5 because the ones digit is 5.

325 ÷ 5 = 65 so 325 = 5·65 and 7800 = 2·2·2·3·5·65.

Since 325 is not divisible by 3, none of its factors is divisible by 3. Therefore we no longer need to check for divisibility by 3.

65 is divisible by 5 because the ones digit is 5.

65 ÷ 5 = 13 so 65 = 5·13 and 7800 = 2·2·2·3·5·5·13.

13 is prime so the prime factorization of 7800 is 2·2·2·3·5·5·13.

23. 2772 is divisible by 2 because the ones digit (2) is even.

2772 ÷ 2 = 1386 so 2772 = 2·1386.

1386 is divisible by 2 because the ones digit (6) is even.

1386 ÷ 2 = 693 so 1386 = 2·693 and 2772 = 2·2·693.

693 is not divisible by 2 because the ones digit (3) is not even. We move to 3.

693 is divisible by 3 because the sum of the digits (6 + 9 + 3 = 18) is divisible by 3.

693 ÷ 3 = 231 so 693 = 3·231 and 2772 = 2·2·3·231.

Since 693 is not divisible by 2, none of its factors is divisible by 2. Therefore, we no longer need to check divisibility by 2.

231 is divisible by 3 because the sum of the digits (2 + 3 + 1 = 6) is divisible by 3.

231 ÷ 3 = 77 so 231 = 3·77 and 2772 = 2·2·3·3·77.

77 is not divisible by 3 since the sum of the digits (7 + 7 = 14) is not divisible by 3. We move to 5.

77 is not divisible by 5 because the ones digit (7) is not 0 or 5. We move to 7.

We have not stated a test for divisibility by 7 so we will just try dividing by 7.

$$7\overline{)77}^{\,11} \quad \longleftarrow \quad \text{11 is prime}$$

77 ÷ 7 = 11 so 77 = 7·11 and the prime factorization of 2772 is 2·2·3·3·7·11.

Exercise Set 2.3

1. The top number is the numerator, and the bottom number is the denominator.

   $$\frac{3}{4} \quad \begin{matrix} \longleftarrow & \text{Numerator} \\ \longleftarrow & \text{Denominator} \end{matrix}$$

3. $$\frac{11}{20} \quad \begin{matrix} \longleftarrow & \text{Numerator} \\ \longleftarrow & \text{Denominator} \end{matrix}$$

5. The dollar is divided into 4 parts of the same size, and 2 of them are shaded. This is $2 \cdot \frac{1}{4}$ or $\frac{2}{4}$. Thus, $\frac{2}{4}$ (two-fourths) of the dollar is shaded.

7. The yard is divided into 8 parts of the same size, and 1 of them is shaded. Thus, $\frac{1}{8}$ (one-eighth) of the yard is shaded.

9. We have 2 quarts, each divided into thirds. We take $\frac{1}{3}$ of each. This is $2 \cdot \frac{1}{3}$ or $\frac{2}{3}$. Thus, $\frac{2}{3}$ of a quart is shaded.

11. The triangle is divided into 4 triangles of the same size, and 3 of them are shaded. This is $3 \cdot \frac{1}{4}$ or $\frac{3}{4}$. Thus, $\frac{3}{4}$ (three-fourths) of the triangle is shaded.

13. The pie is divided into 8 parts of the same size, and 4 of them are shaded. This is $4 \cdot \frac{1}{8}$, or $\frac{4}{8}$. Thus, $\frac{4}{8}$ (four-eighths) of the pie is shaded.

15. The acre is divided into 12 parts of the same size, and 6 of them are shaded. This is $6 \cdot \frac{1}{12}$, or $\frac{6}{12}$. Thus, $\frac{6}{12}$ (six-twelfths) of the acre is shaded.

17. There are 8 circles, and 5 are shaded. Thus, $\frac{5}{8}$ of the circles are shaded.

19. There are 5 objects in the set, and 3 of the objects are shaded. Thus, $\frac{3}{5}$ of the set is shaded.

21. Remember: $\frac{0}{n} = 0$, for n that is not 0.

    $\frac{0}{5} = 0$

    Think of dividing an object into 5 parts and taking none of them. We get 0.

23. Remember: $\frac{n}{1} = n$.

$\frac{5}{1} = 5$

Think of taking 5 objects and dividing them into 1 part. (We do not divide them.) We have 5 objects.

25. Remember: $\frac{n}{n} = 1$, for n that is not 0.

$\frac{20}{20} = 1$

If we divide an object into 20 parts and take 20 of them, we get all of the object (1 whole object).

27. Remember: $\frac{n}{n} = 1$, for n that is not 0.

$\frac{45}{45} = 1$

If we divide an object into 45 parts, and take 45 of them, we get all of the object (1 whole object).

29. Remember: $\frac{0}{n} = 0$, for n that is not 0.

$\frac{0}{234} = 0$

Think of dividing an object into 234 parts and taking none of them. We get 0.

31. Remember: $\frac{n}{n} = 1$, for n that is not 0.

$\frac{234}{234} = 1$

If we divide an object into 234 parts and take 234 of them, we get all of the object (1 whole object).

33. Remember: $\frac{n}{n} = 1$, for n that is not 0.

$\frac{3}{3} = 1$

If we divide an object into 3 parts and take 3 of them, we get all of the object (1 whole object).

35. Remember: $\frac{n}{n} = 1$, for n that is not 0.

$\frac{57}{57} = 1$

If we divide an object into 57 parts and take 57 of them, we get all of the object (1 whole object).

37. Remember: $\frac{n}{n} = 1$, for n that is not 0.

$\frac{8}{8} = 1$

39. Remember: $\frac{n}{1} = n$

$\frac{8}{1} = 8$

41. Round 34,56 $\boxed{2}$ to the nearest ten.

The digit 6 is in the tens place. Consider the next digit to the right. Since 2 is 4 or lower, we round down.

The answer is 34,560.

43. Round 34, $\boxed{5}$ 62 to the nearest thousand.

The digit 4 is in the thousands place. Consider the next digit to the right. Since 5 is 5 or higher, we round up.

The answer is 35,000.

45. Familiarize. We visualize the situation.

62 words	62 words	$\cdots$	62 words
12,462 words			

Translate. We let t = the time it will take to type 12,462 words.

Number of words per minute	times	Number of minutes	is	Total number of words
62	$\cdot$	t	=	12,462

Solve. We solve the equation.

$62 \cdot t = 12,462$

$\frac{62 \cdot t}{62} = \frac{12,462}{62}$

$t = 201$

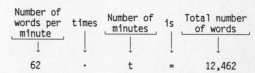

Check. We can multiply 62 by the number of minutes: $201 \cdot 62 = 12,462$. The answer checks.

State. The answer is 201 minutes.

47. Think of dividing $2700 into 2700 parts of the same size. (Each part is $1.)

a) Think of $1200 as 1200 of the 2700 parts. Thus, we have $1200 \cdot \frac{1}{2700}$ or $\frac{1200}{2700}$.

b) Think of $540 as 540 of the 2700 parts. Thus, we have $540 \cdot \frac{1}{2700}$ or $\frac{540}{2700}$.

c) We have $360 \cdot \frac{1}{2700}$ or $\frac{360}{2700}$.

d) First, find the amount of miscellaneous expenses. To do this we must find the total of the amounts spent for tuition, rent, and food and then take that amount away from the amount earned.

47. (continued)

To find the total spent for tuition, rent, and food translate to an equation and solve.

$$1200 + 540 + 360 = x$$
$$2100 = x \quad \text{Doing the addition}$$

The total is $2100.

Now we have a "take-away" situation:

$$2700 - 2100 = m$$
$$600 = m \quad \text{Doing the subtraction}$$

Thus, $600 went for miscellaneous expenses.

Think of $600 as 600 of the 2700 parts. Thus, we have $600 \cdot \frac{1}{2700} = \frac{600}{2700}$.

Exercise Set 2.4

1. $3 \cdot \frac{1}{5} = \frac{3 \cdot 1}{5} = \frac{3}{5}$

3. $5 \times \frac{1}{6} = \frac{5 \times 1}{6} = \frac{5}{6}$

5. $\frac{2}{11} \cdot 4 = \frac{2 \cdot 4}{11} = \frac{8}{11}$

7. $10 \cdot \frac{7}{9} = \frac{10 \cdot 7}{9} = \frac{70}{9}$

9. $\frac{2}{5} \cdot 1 = \frac{2 \cdot 1}{5} = \frac{2}{5}$

11. $\frac{2}{5} \cdot 3 = \frac{2 \cdot 3}{5} = \frac{6}{5}$

13. $7 \cdot \frac{3}{4} = \frac{7 \cdot 3}{4} = \frac{21}{4}$

15. $17 \times \frac{5}{6} = \frac{17 \times 5}{6} = \frac{85}{6}$

17. $\frac{1}{2} \cdot \frac{1}{3} = \frac{1 \cdot 1}{2 \cdot 3} = \frac{1}{6}$

19. $\frac{1}{4} \times \frac{1}{10} = \frac{1 \times 1}{4 \times 10} = \frac{1}{40}$

21. $\frac{2}{3} \times \frac{1}{5} = \frac{2 \times 1}{3 \times 5} = \frac{2}{15}$

23. $\frac{2}{5} \cdot \frac{2}{3} = \frac{2 \cdot 2}{5 \cdot 3} = \frac{4}{15}$

25. $\frac{3}{4} \cdot \frac{3}{4} = \frac{3 \cdot 3}{4 \cdot 4} = \frac{9}{16}$

27. $\frac{2}{3} \cdot \frac{7}{13} = \frac{2 \cdot 7}{3 \cdot 13} = \frac{14}{39}$

29. $\frac{1}{10} \cdot \frac{7}{10} = \frac{1 \cdot 7}{10 \cdot 10} = \frac{7}{100}$

31. $\frac{7}{8} \cdot \frac{7}{8} = \frac{7 \cdot 7}{8 \cdot 8} = \frac{49}{64}$

33. $\frac{1}{10} \cdot \frac{1}{100} = \frac{1 \cdot 1}{10 \cdot 100} = \frac{1}{1000}$

35. $\frac{14}{15} \cdot \frac{13}{19} = \frac{14 \cdot 13}{15 \cdot 19} = \frac{182}{285}$

37. Familiarize. Recall the area is length times width. We draw a picture. We will let A = the area of the table top.

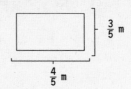

Translate. Then we translate.

Area is length times width
$$A = \frac{4}{5} \times \frac{3}{5}$$

Solve. The sentence tells us what to do. We multiply.

$$\frac{4}{5} \times \frac{3}{5} = \frac{4 \times 3}{5 \times 5} = \frac{12}{25}$$

Check. We repeat the calculation. The answer checks.

State. The area is $\frac{12}{25}$ square meter.

39. Familiarize. We draw a picture. We let n = the amount the can holds when it is $\frac{1}{2}$ full.

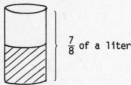

$\frac{1}{2}$ of $\frac{7}{8}$ liter

Translate. The multiplication sentence $\frac{1}{2} \cdot \frac{7}{8} = n$ corresponds to the situation.

Solve. We multiply:

$$\frac{1}{2} \cdot \frac{7}{8} = \frac{1 \cdot 7}{2 \cdot 8} = \frac{7}{16}$$

Check. We repeat the calculation. The answer checks.

State. The can will hold $\frac{7}{16}$ of a liter when it is full.

41. <u>Familiarize</u>. We draw a picture. We let h = the amount of honey that is needed.

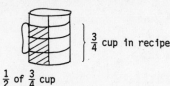

$\frac{1}{2}$ of $\frac{3}{4}$ cup

$\left.\begin{array}{c}\\\end{array}\right\}$ $\frac{3}{4}$ cup in recipe

<u>Translate</u>. The multiplication sentence $\frac{1}{2} \cdot \frac{3}{4}$ = h corresponds to the situation.

<u>Solve</u>. We multiply:

$\frac{1}{2} \cdot \frac{3}{4} = \frac{1 \cdot 3}{2 \cdot 4} = \frac{3}{8}$

<u>Check</u>. We repeat the calculation. The answer checks.

<u>State</u>. $\frac{3}{8}$ cup of honey is needed.

43. <u>Familiarize</u>. Think of each person as $\frac{1}{1000}$ of the of the entire group. We let m = the fraction in the 16-20 age group.

<u>Translate</u>. The multiplication sentence $230 \cdot \frac{1}{1000}$ = m corresponds to the situation.

<u>Solve</u>. We multiply:

$230 \cdot \frac{1}{1000} = \frac{230 \cdot 1}{1000} = \frac{230}{1000}$

<u>Check</u>. We repeat the calculation. The answer checks.

<u>State</u>. $\frac{230}{1000}$ of all moviegoers are in the 16-20 age group.

45.
```
       2 0 4
3 5 ) 7 1 4 0
      7 0 0 0
        1 4 0
        1 4 0
            0
```

The answer is 204.

47.
```
        3 0 0 1
9 ) 2 7,0 0 9
    2 7 0 0 0
            9
            9
            0
```

The answer is 3001.

<u>Exercise Set 2.5</u>

1. Since 2·5 = 10, we multiply by $\frac{5}{5}$.

$\frac{1}{2} = \frac{1}{2} \cdot \frac{5}{5} = \frac{1 \cdot 5}{2 \cdot 5} = \frac{5}{10}$

3. Since 4·12 = 48, we multiply by $\frac{12}{12}$.

$\frac{3}{4} = \frac{3}{4} \cdot \frac{12}{12} = \frac{3 \cdot 12}{4 \cdot 12} = \frac{36}{48}$

5. Since 10·3 = 30, we multiply by $\frac{3}{3}$.

$\frac{9}{10} = \frac{9}{10} \cdot \frac{3}{3} = \frac{9 \cdot 3}{10 \cdot 3} = \frac{27}{30}$

7. Since 8·4 = 32, we multiply by $\frac{4}{4}$.

$\frac{7}{8} = \frac{7}{8} \cdot \frac{4}{4} = \frac{7 \cdot 4}{8 \cdot 4} = \frac{28}{32}$

9. Since 12·4 = 48, we multiply by $\frac{4}{4}$.

$\frac{5}{12} = \frac{5}{12} \cdot \frac{4}{4} = \frac{5 \cdot 4}{12 \cdot 4} = \frac{20}{48}$

11. Since 18·3 = 54, we multiply by $\frac{3}{3}$.

$\frac{17}{18} = \frac{17}{18} \cdot \frac{3}{3} = \frac{17 \cdot 3}{18 \cdot 3} = \frac{51}{54}$

13. Since 3·15 = 45, we multiply by $\frac{15}{15}$.

$\frac{5}{3} = \frac{5}{3} \cdot \frac{15}{15} = \frac{5 \cdot 15}{3 \cdot 15} = \frac{75}{45}$

15. Since 22·6 = 132, we multiply by $\frac{6}{6}$.

$\frac{7}{22} = \frac{7}{22} \cdot \frac{6}{6} = \frac{7 \cdot 6}{22 \cdot 6} = \frac{42}{132}$

17. $\frac{2}{4} = \frac{1 \cdot 2}{2 \cdot 2}$ ⟵ Factor the numerator
      ⟵ Factor the denominator

$= \frac{1}{2} \cdot \frac{2}{2}$ ⟵ Factor the fraction

$= \frac{1}{2} \cdot 1$ ⟵ $\frac{2}{2} = 1$

$= \frac{1}{2}$ ⟵ Removing a factor of 1

19. $\frac{6}{8} = \frac{3 \cdot 2}{4 \cdot 2}$ ⟵ Factor the numerator
      ⟵ Factor the denominator

$= \frac{3}{4} \cdot \frac{2}{2}$ ⟵ Factor the fraction

$= \frac{3}{4} \cdot 1$ ⟵ $\frac{2}{2} = 1$

$= \frac{3}{4}$ ⟵ Removing a factor of 1

21. $\dfrac{3}{15} = \dfrac{1 \cdot 3}{5 \cdot 3}$ ⟵ Factor the numerator
   ⟵ Factor the denominator

   $= \dfrac{1}{5} \cdot \dfrac{3}{3}$ ⟵ Factor the fraction

   $= \dfrac{1}{5} \cdot 1$ ⟵ $\dfrac{3}{3} = 1$

   $= \dfrac{1}{5}$ ⟵ Removing a factor of 1

23. $\dfrac{24}{8} = \dfrac{3 \cdot 8}{1 \cdot 8} = \dfrac{3}{1} \cdot \dfrac{8}{8} = \dfrac{3}{1} \cdot 1 = \dfrac{3}{1} = 3$

25. $\dfrac{18}{24} = \dfrac{3 \cdot 6}{4 \cdot 6} = \dfrac{3}{4} \cdot \dfrac{6}{6} = \dfrac{3}{4} \cdot 1 = \dfrac{3}{4}$

27. $\dfrac{14}{16} = \dfrac{7 \cdot 2}{8 \cdot 2} = \dfrac{7}{8} \cdot \dfrac{2}{2} = \dfrac{7}{8} \cdot 1 = \dfrac{7}{8}$

29. $\dfrac{12}{10} = \dfrac{6 \cdot 2}{5 \cdot 2} = \dfrac{6}{5} \cdot \dfrac{2}{2} = \dfrac{6}{5} \cdot 1 = \dfrac{6}{5}$

31. $\dfrac{16}{48} = \dfrac{1 \cdot 16}{3 \cdot 16} = \dfrac{1}{3} \cdot \dfrac{16}{16} = \dfrac{1}{3} \cdot 1 = \dfrac{1}{3}$

33. $\dfrac{150}{25} = \dfrac{6 \cdot 25}{1 \cdot 25} = \dfrac{6}{1} \cdot \dfrac{25}{25} = \dfrac{6}{1} \cdot 1 = \dfrac{6}{1} = 6$

   We could also simplify $\dfrac{150}{25}$ by doing the division $150 \div 25$. That is, $\dfrac{150}{25} = 150 \div 25 = 6$.

35. $\dfrac{17}{51} = \dfrac{1 \cdot 17}{3 \cdot 17} = \dfrac{1}{3} \cdot \dfrac{17}{17} = \dfrac{1}{3} \cdot 1 = \dfrac{1}{3}$

37. We multiply these   We multiply these
   two numbers: two numbers:

   $3 \cdot 12 = 36$ $4 \cdot 9 = 36$

   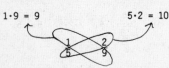

   Since $36 = 36$, $\dfrac{3}{4} = \dfrac{9}{12}$.

39. We multiply these   We multiply these
   two numbers: two numbers:

   $1 \cdot 9 = 9$ $5 \cdot 2 = 10$

   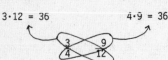

   Since $9 \neq 10$, $\dfrac{1}{5} \neq \dfrac{2}{9}$.

41. We multiply these   We multiply these
   two numbers: two numbers:

   $3 \cdot 16 = 48$ $8 \cdot 6 = 48$

   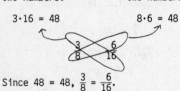

   Since $48 = 48$, $\dfrac{3}{8} = \dfrac{6}{16}$.

43. We multiply these   We multiply these
   two numbers: two numbers:

   $2 \cdot 7 = 14$ $5 \cdot 3 = 15$

   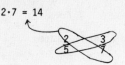

   Since $14 \neq 15$, $\dfrac{2}{5} \neq \dfrac{3}{7}$.

45. We multiply these   We multiply these
   two numbers: two numbers:

   $12 \cdot 6 = 72$ $9 \cdot 8 = 72$

   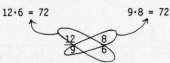

   Since $72 = 72$, $\dfrac{12}{9} = \dfrac{8}{6}$.

47. We multiply these   We multiply these
   two numbers: two numbers:

   $5 \cdot 7 = 35$ $2 \cdot 17 = 34$

   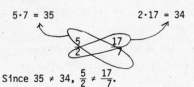

   Since $35 \neq 34$, $\dfrac{5}{2} \neq \dfrac{17}{7}$.

49. We multiply these   We multiply these
   two numbers: two numbers:

   $3 \cdot 100 = 300$ $10 \cdot 30 = 300$

   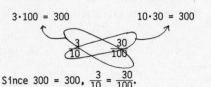

   Since $300 = 300$, $\dfrac{3}{10} = \dfrac{30}{100}$.

51. We multiply these   We multiply these
   two numbers: two numbers:

   $5 \cdot 1000 = 5000$ $10 \cdot 520 = 5200$

   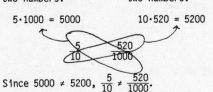

   Since $5000 \neq 5200$, $\dfrac{5}{10} \neq \dfrac{520}{1000}$.

53. <u>Familiarize</u>. We make a drawing. We let A = the area.

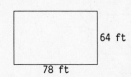

64 ft

78 ft

<u>Translate</u>. Using the formula for area, we have

A = ℓ·w = 78·64.

<u>Solve</u>. We carry out the multiplication.

```
    7 8
  × 6 4
    3 1 2
  4 6 8 0
  4 9 9 2
```

Thus, A = 4992.

<u>Check</u>. We repeat the calculation. The answer checks.

<u>State</u>. The area is 4992 square feet.

55. Think of each person as $\frac{1}{10}$. Then the multiplication sentence $4 \cdot \frac{1}{10} = n$ corresponds to the situation. We multiply:

$$4 \cdot \frac{1}{10} = \frac{4 \cdot 1}{10} = \frac{4}{10}$$

Then we simplify:

$$\frac{4}{10} = \frac{2 \cdot 2}{5 \cdot 2} = \frac{2}{5} \cdot \frac{2}{2} = \frac{2}{5} \cdot 1 = \frac{2}{5}$$

57. The batting average is a fraction with the number of hits as its numerator and the number of at bats as its denominator. Jim Rice's batting average was $\frac{92}{564}$. George Brett's batting average was $\frac{84}{634}$. Test these fractions for equality:

We multiply these two numbers (using a calculator):

We multiply these two numbers (using a calculator):

92·634 = 58,328   564·84 = 47,376

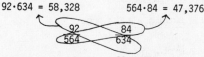

Since 58,328 ≠ 47,376, $\frac{92}{564} \neq \frac{84}{634}$ and the batting averages are not the same.

Exercise Set 2.6

1. $\frac{2}{3} \cdot \frac{1}{2} = \frac{2 \cdot 1}{3 \cdot 2} = \frac{2}{2} \cdot \frac{1}{3} = \frac{1}{3}$

3. $\frac{7}{8} \cdot \frac{1}{7} = \frac{7 \cdot 1}{8 \cdot 7} = \frac{7}{7} \cdot \frac{1}{8} = \frac{1}{8}$

5. $\frac{1}{8} \cdot \frac{4}{5} = \frac{1 \cdot 4}{8 \cdot 5} = \frac{1 \cdot 4}{2 \cdot 4 \cdot 5} = \frac{4}{4} \cdot \frac{1}{2 \cdot 5} = \frac{1}{2 \cdot 5} = \frac{1}{10}$

7. $\frac{1}{4} \cdot \frac{2}{3} = \frac{1 \cdot 2}{4 \cdot 3} = \frac{1 \cdot 2}{2 \cdot 2 \cdot 3} = \frac{2}{2} \cdot \frac{1}{2 \cdot 3} = \frac{1}{2 \cdot 3} = \frac{1}{6}$

9. $\frac{12}{5} \cdot \frac{9}{8} = \frac{12 \cdot 9}{5 \cdot 8} = \frac{4 \cdot 3 \cdot 9}{5 \cdot 2 \cdot 4} = \frac{4}{4} \cdot \frac{3 \cdot 9}{5 \cdot 2} = \frac{3 \cdot 9}{5 \cdot 2} = \frac{27}{10}$

11. $\frac{10}{9} \cdot \frac{7}{5} = \frac{10 \cdot 7}{9 \cdot 5} = \frac{5 \cdot 2 \cdot 7}{9 \cdot 5} = \frac{5}{5} \cdot \frac{2 \cdot 7}{9} = \frac{2 \cdot 7}{9} = \frac{14}{9}$

13. $9 \cdot \frac{1}{9} = \frac{9 \cdot 1}{9} = \frac{9 \cdot 1}{9 \cdot 1} = 1$

15. $\frac{1}{3} \cdot 3 = \frac{1 \cdot 3}{3} = \frac{1 \cdot 3}{1 \cdot 3} = 1$

17. $\frac{7}{10} \cdot \frac{10}{7} = \frac{7 \cdot 10}{10 \cdot 7} = \frac{7 \cdot 10}{7 \cdot 10} = 1$

19. $\frac{7}{5} \cdot \frac{5}{7} = \frac{7 \cdot 5}{5 \cdot 7} = \frac{7 \cdot 5}{7 \cdot 5} = 1$

21. $\frac{1}{4} \cdot 8 = \frac{1 \cdot 8}{4} = \frac{8}{4} = \frac{4 \cdot 2}{4 \cdot 1} = \frac{4}{4} \cdot \frac{2}{1} = \frac{2}{1} = 2$

23. $15 \cdot \frac{1}{3} = \frac{15 \cdot 1}{3} = \frac{15}{3} = \frac{3 \cdot 5}{3 \cdot 1} = \frac{3}{3} \cdot \frac{5}{1} = \frac{5}{1} = 5$

25. $12 \cdot \frac{3}{4} = \frac{12 \cdot 3}{4} = \frac{4 \cdot 3 \cdot 3}{4 \cdot 1} = \frac{4}{4} \cdot \frac{3 \cdot 3}{1} = \frac{3 \cdot 3}{1} = 9$

27. $\frac{3}{8} \cdot 24 = \frac{3 \cdot 24}{8} = \frac{3 \cdot 3 \cdot 8}{1 \cdot 8} = \frac{8}{8} \cdot \frac{3 \cdot 3}{1} = \frac{3 \cdot 3}{1} = 9$

29. $13 \cdot \frac{2}{5} = \frac{13 \cdot 2}{5} = \frac{26}{5}$

31. $\frac{7}{10} \cdot 28 = \frac{7 \cdot 28}{10} = \frac{7 \cdot 2 \cdot 14}{2 \cdot 5} = \frac{2}{2} \cdot \frac{7 \cdot 14}{5} = \frac{7 \cdot 14}{5} = \frac{98}{5}$

33. $\frac{1}{6} \cdot 360 = \frac{1 \cdot 360}{6} = \frac{360}{6} = \frac{6 \cdot 60}{6 \cdot 1} = \frac{6}{6} \cdot \frac{60}{1} = \frac{60}{1} = 60$

35. $240 \cdot \frac{1}{8} = \frac{240 \cdot 1}{8} = \frac{240}{8} = \frac{8 \cdot 30}{8 \cdot 1} = \frac{8}{8} \cdot \frac{30}{1} = \frac{30}{1} = 30$

37. $\frac{4}{10} \cdot \frac{5}{10} = \frac{4 \cdot 5}{10 \cdot 10} = \frac{2 \cdot 2 \cdot 5 \cdot 1}{2 \cdot 5 \cdot 2 \cdot 5} = \frac{2 \cdot 2 \cdot 5}{2 \cdot 2 \cdot 5} \cdot \frac{1}{5} = \frac{1}{5}$

39. $\frac{8}{10} \cdot \frac{45}{100} = \frac{8 \cdot 45}{10 \cdot 100} = \frac{2 \cdot 2 \cdot 2 \cdot 5 \cdot 9}{2 \cdot 5 \cdot 2 \cdot 5 \cdot 2 \cdot 5} = \frac{2 \cdot 2 \cdot 2 \cdot 5}{2 \cdot 2 \cdot 2 \cdot 5} \cdot \frac{9}{5 \cdot 5}$
$= \frac{9}{5 \cdot 5} = \frac{9}{25}$

41. $\frac{11}{24} \cdot \frac{3}{5} = \frac{11 \cdot 3}{24 \cdot 5} = \frac{11 \cdot 3}{3 \cdot 8 \cdot 5} = \frac{3}{3} \cdot \frac{11}{8 \cdot 5} = \frac{11}{8 \cdot 5} = \frac{11}{40}$

43. $\frac{10}{21} \cdot \frac{3}{4} = \frac{10 \cdot 3}{21 \cdot 4} = \frac{2 \cdot 5 \cdot 3}{3 \cdot 7 \cdot 2 \cdot 2} = \frac{2 \cdot 3}{2 \cdot 3} \cdot \frac{5}{7 \cdot 2} = \frac{5}{7 \cdot 2} = \frac{5}{14}$

45. Familiarize. We visualize the situation. We let a = the amount received for working $\frac{1}{4}$ of a day.

1 day $36			
1/4 day $a			

Translate. We write an equation.

$$\begin{array}{cccc} \text{Pay for} & \text{is} & \frac{1}{4} & \text{of} & \$36 \\ \text{1/4 of a day,} & & & & \\ \downarrow & \downarrow & \downarrow & \downarrow & \downarrow \\ a & = & \frac{1}{4} & \cdot & 36 \end{array}$$

Solve. We carry out the multiplication.

$$a = \frac{1}{4} \cdot 36 = \frac{1 \cdot 36}{4} = \frac{36}{4}$$

$$= \frac{9 \cdot 4}{1 \cdot 4} = \frac{9}{1} \cdot \frac{4}{4}$$

$$= 9$$

Check. We can repeat the calculation. We can also determine that the answer seems reasonable since we multiplied 36 by a number less than 1 and the result is less than 36. The answer checks.

State. $9 is received for working $\frac{1}{4}$ of a day.

47. Familiarize. We visualize the situation. We let n = the number of addresses that will be incorrect after one year.

Mailing list 2500 addresses			
1/4 of the addresses n			

Translate.

$$\begin{array}{cccc} \text{Number} & \text{is} & \frac{1}{4} & \text{of} & \text{Number} \\ \text{incorrect,} & & & & \text{of addresses,} \\ \downarrow & \downarrow & \downarrow & \downarrow & \downarrow \\ n & = & \frac{1}{4} & \cdot & 2500 \end{array}$$

Solve. We carry out the multiplication.

$$n = \frac{1}{4} \cdot 2500 = \frac{1 \cdot 2500}{4} = \frac{2500}{4}$$

$$= \frac{4 \cdot 625}{4 \cdot 1} = \frac{4}{4} \cdot \frac{625}{1}$$

$$= 625$$

Check. We can repeat the calculation. We can also determine that the answer seems reasonable since we multiplied 2500 by a number less than 1 and the result is less than 2500. The answer checks.

State. After one year 625 addresses will be incorrect.

49. Familiarize. We draw a picture.

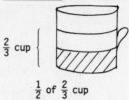

$\frac{2}{3}$ cup

$\frac{1}{2}$ of $\frac{2}{3}$ cup

We let n = the amount of flour the chef should use.

Translate. The multiplication sentence

$$\frac{1}{2} \cdot \frac{2}{3} = n$$

corresponds to the situation.

Solve. We multiply and simplify:

$$n = \frac{1}{2} \cdot \frac{2}{3} = \frac{1 \cdot 2}{2 \cdot 3} = \frac{2}{2} \cdot \frac{1}{3} = \frac{1}{3}$$

Check. We can repeat the calculation. We can also determine that the answer seems reasonable since we multiplied $\frac{2}{3}$ by a number less than 1 and the result is less than $\frac{2}{3}$. The answer checks.

State. The chef should use $\frac{1}{3}$ cup of flour.

51. Familiarize. We visualize the situation. Let a = the amount of the loan.

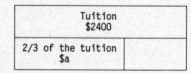

Tuition $2400	
2/3 of the tuition $a	

Translate. We write an equation.

$$\begin{array}{cccc} \text{Amount} & \text{is} & \frac{2}{3} & \text{of} & \text{the tuition,} \\ \text{of loan,} & & & & \\ \downarrow & \downarrow & \downarrow & \downarrow & \downarrow \\ a & = & \frac{2}{3} & \cdot & 2400 \end{array}$$

Solve. We carry out the multiplication.

$$a = \frac{2}{3} \cdot 2400 = \frac{2 \cdot 2400}{3}$$

$$= \frac{2 \cdot 3 \cdot 800}{3 \cdot 1} = \frac{3}{3} \cdot \frac{2 \cdot 800}{1}$$

$$= \frac{1600}{1} = 1600$$

Check. We can repeat the calculation. We can also determine that the answer seems reasonable since we multiplied 2400 by a number less than 1 and the result is less than 2400. The answer checks.

State. The loan was $1600.

53. <u>Familiarize</u>. We draw a picture.

$\frac{2}{3}$ in.

1 in.

240 miles

We let n = the number of miles represented by $\frac{2}{3}$ in.

<u>Translate</u>. The multiplication sentence

$n = \frac{2}{3} \cdot 240$

corresponds to the situation.

<u>Solve</u>. We multiply and simplify:

$n = \frac{2}{3} \cdot 240 = \frac{2 \cdot 240}{3} = \frac{2 \cdot 3 \cdot 80}{1 \cdot 3}$

$= \frac{3}{3} \cdot \frac{2 \cdot 80}{1} = \frac{2 \cdot 80}{1}$

$= \frac{160}{1} = 160$

<u>Check</u>. We can repeat the calculation. We can also determine that the answer seems reasonable since we multiplied 240 by a number less than 1 and the result is less than 240.

<u>State</u>. $\frac{2}{3}$ in. on the map represents 160 miles.

55. <u>Familiarize</u>. This is a multistep problem. First we find the amount of each of the given expenses. Then we find the total of these expenses and take it away from the annual income to find how much is spent for other expenses.

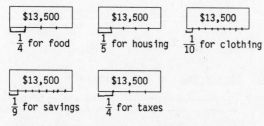

$\frac{1}{4}$ for food $\frac{1}{5}$ for housing $\frac{1}{10}$ for clothing

$\frac{1}{9}$ for savings $\frac{1}{4}$ for taxes

We let f, h, c, s, and t represent the amounts spent on food, housing, clothing, savings, and taxes, respectively.

<u>Translate</u>. The following multiplication sentences correspond to the situation.

$\frac{1}{4} \cdot 13{,}500 = f$ $\frac{1}{9} \cdot 13{,}500 = s$

$\frac{1}{5} \cdot 13{,}500 = h$ $\frac{1}{4} \cdot 13{,}500 = t$

$\frac{1}{10} \cdot 13{,}500 = c$

55. (continued)

<u>Solve</u>. We multiply and simplify each situation:

$f = \frac{1}{4} \cdot 13{,}500 = \frac{13{,}500}{4} = \frac{4 \cdot 3375}{1 \cdot 4} = \frac{4}{4} \cdot \frac{3375}{1} = \frac{3375}{1}$
$= 3375$

$h = \frac{1}{5} \cdot 13{,}500 = \frac{13{,}500}{5} = \frac{5 \cdot 2700}{1 \cdot 5} = \frac{5}{5} \cdot \frac{2700}{1} = \frac{2700}{1}$
$= 2700$

$c = \frac{1}{10} \cdot 13{,}500 = \frac{13{,}500}{10} = \frac{10 \cdot 1350}{1 \cdot 10} = \frac{10}{10} \cdot \frac{1350}{1}$
$= \frac{1350}{1} = 1350$

$s = \frac{1}{9} \cdot 13{,}500 = \frac{13{,}500}{9} = \frac{9 \cdot 1500}{1 \cdot 9} = \frac{9}{9} \cdot \frac{1500}{1} = \frac{1500}{1}$
$= 1500$

$t = \frac{1}{4} \cdot 13{,}500 = \frac{13{,}500}{4} = \frac{4 \cdot 3375}{1 \cdot 4} = \frac{4}{4} \cdot \frac{3375}{1} = \frac{3375}{1}$
$= 3375$

We add to find the total of these expenses.

$$
\begin{array}{r}
\$\ 3\ 3\ 7\ 5 \\
2\ 7\ 0\ 0 \\
1\ 3\ 5\ 0 \\
1\ 5\ 0\ 0 \\
\underline{3\ 3\ 7\ 5} \\
\$\ 1\ 2{,}3\ 0\ 0
\end{array}
$$

We let m = the amount spent on other expenses and subtract to find this amount.

Annual income	minus	Total of itemized expenses	is	Total spent on other expenses
↓	↓	↓	↓	↓
$13,500	−	$12,300	=	m

$1200 = m   Subtracting

<u>Check</u>. We repeat the calculations. The results check.

<u>State</u>. $3375 is spent for food, $2700 for housing, $1350 for clothing, $1500 for savings, $3375 for taxes, and $1200 for other expenses.

57. $48 \cdot t = 1680$

$\frac{48 \cdot t}{48} = \frac{1680}{48}$

$t = 35$

The solution is 35.

59.

$$
\begin{array}{r}
9\ 0\ 6\ 0 \\
-\ 4\ 3\ 8\ 7 \\
\hline
4\ 6\ 7\ 3
\end{array}
$$

Exercise Set 2.7

1. $\dfrac{5}{6}\searrow\nearrow\dfrac{6}{5}$ Interchange the numerator and denominator

The reciprocal of $\dfrac{5}{6}$ is $\dfrac{6}{5}$.

3. Think of 6 as $\dfrac{6}{1}$.

$\dfrac{6}{1}\searrow\nearrow\dfrac{1}{6}$

The reciprocal of 6 is $\dfrac{1}{6}$.

5. $\dfrac{1}{6}\searrow\nearrow\dfrac{6}{1}$ $\left(\dfrac{6}{1}=6\right)$

The reciprocal of $\dfrac{1}{6}$ is 6.

7. $\dfrac{10}{3}\searrow\nearrow\dfrac{3}{10}$

The reciprocal of $\dfrac{10}{3}$ is $\dfrac{3}{10}$.

9. $\dfrac{3}{5}\div\dfrac{3}{4}=\dfrac{3}{5}\cdot\dfrac{4}{3}$ Multiplying the dividend $\left(\dfrac{3}{5}\right)$ by the reciprocal of the divisor $\left[\text{The reciprocal of }\dfrac{3}{4}\text{ is }\dfrac{4}{3}.\right]$

$=\dfrac{3\cdot4}{5\cdot3}$ Multiplying numerators and denominators

$=\dfrac{3}{3}\cdot\dfrac{4}{5}=\dfrac{4}{5}$ Simplifying

11. $\dfrac{3}{5}\div\dfrac{9}{4}=\dfrac{3}{5}\cdot\dfrac{4}{9}$ Multiplying the dividend $\left(\dfrac{3}{5}\right)$ by the reciprocal of the divisor $\left[\text{The reciprocal of }\dfrac{9}{4}\text{ is }\dfrac{4}{9}.\right]$

$=\dfrac{3\cdot4}{5\cdot9}$ Multiplying numerators and denominators

$=\dfrac{3\cdot4}{5\cdot3\cdot3}=\dfrac{3}{3}\cdot\dfrac{4}{5\cdot3}$ Simplifying

$=\dfrac{4}{5\cdot3}=\dfrac{4}{15}$

13. $\dfrac{4}{3}\div\dfrac{1}{3}=\dfrac{4}{3}\cdot3=\dfrac{4\cdot3}{3}=\dfrac{3}{3}\cdot4=4$

15. $\dfrac{1}{3}\div\dfrac{1}{6}=\dfrac{1}{3}\cdot6=\dfrac{1\cdot6}{3}=\dfrac{1\cdot2\cdot3}{1\cdot3}=\dfrac{1\cdot3}{1\cdot3}\cdot2=2$

17. $\dfrac{3}{8}\div3=\dfrac{3}{8}\cdot\dfrac{1}{3}=\dfrac{3\cdot1}{8\cdot3}=\dfrac{3}{3}\cdot\dfrac{1}{8}=\dfrac{1}{8}$

19. $\dfrac{12}{7}\div4=\dfrac{12}{7}\cdot\dfrac{1}{4}=\dfrac{12\cdot1}{7\cdot4}=\dfrac{4\cdot3\cdot1}{7\cdot4}=\dfrac{4}{4}\cdot\dfrac{3\cdot1}{7}=\dfrac{3\cdot1}{7}=\dfrac{3}{7}$

21. $12\div\dfrac{3}{2}=12\cdot\dfrac{2}{3}=\dfrac{12\cdot2}{3}=\dfrac{3\cdot4\cdot2}{3\cdot1}=\dfrac{3}{3}\cdot\dfrac{4\cdot2}{1}=\dfrac{4\cdot2}{1}$

$=\dfrac{8}{1}=8$

23. $28\div\dfrac{4}{5}=28\cdot\dfrac{5}{4}=\dfrac{28\cdot5}{4}=\dfrac{4\cdot7\cdot5}{4\cdot1}=\dfrac{4}{4}\cdot\dfrac{7\cdot5}{1}=\dfrac{7\cdot5}{1}$

$=35$

25. $\dfrac{5}{8}\div\dfrac{5}{8}=\dfrac{5}{8}\cdot\dfrac{8}{5}=\dfrac{5\cdot8}{8\cdot5}=\dfrac{5\cdot8}{5\cdot8}=1$

27. $\dfrac{8}{15}\div\dfrac{4}{5}=\dfrac{8}{15}\cdot\dfrac{5}{4}=\dfrac{8\cdot5}{15\cdot4}=\dfrac{2\cdot4\cdot5}{3\cdot5\cdot4}=\dfrac{4\cdot5}{4\cdot5}\cdot\dfrac{2}{3}=\dfrac{2}{3}$

29. $\dfrac{9}{5}\div\dfrac{4}{5}=\dfrac{9}{5}\cdot\dfrac{5}{4}=\dfrac{9\cdot5}{5\cdot4}=\dfrac{5}{5}\cdot\dfrac{9}{4}=\dfrac{9}{4}$

31. $120\div\dfrac{5}{6}=120\cdot\dfrac{6}{5}=\dfrac{120\cdot6}{5}=\dfrac{5\cdot24\cdot6}{5\cdot1}=\dfrac{5}{5}\cdot\dfrac{24\cdot6}{1}$

$=\dfrac{24\cdot6}{1}=144$

33. $\dfrac{4}{5}\cdot x=60$

$x=60\div\dfrac{4}{5}$ Dividing on both sides by $\dfrac{4}{5}$

$x=60\cdot\dfrac{5}{4}$ Multiplying by the reciprocal

$=\dfrac{60\cdot5}{4}=\dfrac{4\cdot15\cdot5}{4\cdot1}=\dfrac{4}{4}\cdot\dfrac{15\cdot5}{1}=\dfrac{15\cdot5}{1}=75$

35. $\dfrac{5}{3}\cdot y=\dfrac{10}{3}$

$y=\dfrac{10}{3}\div\dfrac{5}{3}$ Dividing on both sides by $\dfrac{5}{3}$

$y=\dfrac{10}{3}\cdot\dfrac{3}{5}$ Multiplying by the reciprocal

$=\dfrac{10\cdot3}{3\cdot5}=\dfrac{2\cdot5\cdot3}{3\cdot5\cdot1}=\dfrac{5\cdot3}{5\cdot3}\cdot\dfrac{2}{1}=\dfrac{2}{1}=2$

37. $x\cdot\dfrac{25}{36}=\dfrac{5}{12}$

$x=\dfrac{5}{12}\div\dfrac{25}{36}=\dfrac{5}{12}\cdot\dfrac{36}{25}=\dfrac{5\cdot36}{12\cdot25}=\dfrac{5\cdot3\cdot12}{12\cdot5\cdot5}$

$=\dfrac{5\cdot12}{5\cdot12}\cdot\dfrac{3}{5}=\dfrac{3}{5}$

39. $n\cdot\dfrac{8}{7}=360$

$n=360\div\dfrac{8}{7}=360\cdot\dfrac{7}{8}=\dfrac{360\cdot7}{8}=\dfrac{8\cdot45\cdot7}{8\cdot1}$

$=\dfrac{8}{8}\cdot\dfrac{45\cdot7}{1}=\dfrac{45\cdot7}{1}=315$

41. **Familiarize.** We draw a picture.

6 pieces of the same length

$\dfrac{3}{5}$ m

We let n = the length of each piece.

Translate. The multiplication that corresponds to the situation is

$6\cdot n=\dfrac{3}{5}$

Solve. We solve the equation by dividing on both sides by 6 and carrying out the division:

$n=\dfrac{3}{5}\div6=\dfrac{3}{5}\cdot\dfrac{1}{6}=\dfrac{3\cdot1}{5\cdot6}=\dfrac{3\cdot1}{5\cdot2\cdot3}=\dfrac{3}{3}\cdot\dfrac{1}{5\cdot2}=$

$\dfrac{1}{5\cdot2}=\dfrac{1}{10}$

Check. We repeat the calculation. The answer checks.

41. (continued)

State. Each piece is $\frac{1}{10}$ m.

43. Familiarize. We draw a picture. We let n = the number of teeshirts that can be made.

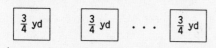

n teeshirts

Translate. The multiplication that corresponds to the situation is

$\frac{3}{4} \cdot n = 24.$

Solve. We solve the equation by dividing on both sides by $\frac{3}{4}$ and carrying out the division:

$n = 24 \div \frac{3}{4} = 24 \cdot \frac{4}{3} = \frac{24 \cdot 4}{3} = \frac{3 \cdot 8 \cdot 4}{3 \cdot 1} = \frac{3}{3} \cdot \frac{8 \cdot 4}{1}$

$= \frac{8 \cdot 4}{1} = 32$

Check. We repeat the calculation. The answer checks.

State. 32 shirts can be made from 24 yd of fabric.

45. Familiarize. We draw a picture. We let n = the number of sugar bowls that can be filled.

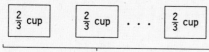

n bowls

Translate. We write a multiplication sentence:

$\frac{2}{3} \cdot n = 16$

Solve. Solve the equation as follows:

$\frac{2}{3} \cdot n = 16$

$n = 16 \div \frac{2}{3} = 16 \cdot \frac{3}{2} = \frac{16 \cdot 3}{2} = \frac{2 \cdot 8 \cdot 3}{2 \cdot 1}$

$= \frac{2}{2} \cdot \frac{8 \cdot 3}{1} = \frac{8 \cdot 3}{1} = 24$

Check. We repeat the calculation. The answer checks.

State. 24 sugar bowls can be filled.

47. Familiarize. We draw a picture. We let n = the amount the bucket could hold.

Translate. We write a multiplication sentence:

$\frac{3}{4} \cdot n = 12$

47. (continued)

Solve. Solve the equation as follows:

$\frac{3}{4} \cdot n = 12$

$n = 12 \div \frac{3}{4} = 12 \cdot \frac{4}{3} = \frac{12 \cdot 4}{3} = \frac{3 \cdot 4 \cdot 4}{3 \cdot 1}$

$= \frac{3}{3} \cdot \frac{4 \cdot 4}{1} = \frac{4 \cdot 4}{1} = 16$

Check. We repeat the calculation. The answer checks.

State. The bucket could hold 16 L.

49. Familiarize. This is a multistep problem. First we find the length of the total trip. Then we find how many kilometers were left to drive. We draw a picture. We let n = the length of the total trip.

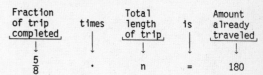

Translate. We translate to an equation.

Fraction of trip completed	times	Total length of trip	is	Amount already traveled
↓	↓	↓	↓	↓
$\frac{5}{8}$	$\cdot$	n	=	180

Solve. We solve the equation as follows:

$\frac{5}{8} \cdot n = 180$

$n = 180 \div \frac{5}{8} = 180 \cdot \frac{8}{5} = \frac{5 \cdot 36 \cdot 8}{5 \cdot 1}$

$= \frac{5}{5} \cdot \frac{36 \cdot 8}{1} = \frac{36 \cdot 8}{1} = 288$

The total trip was 288 km.

Now we find how many kilometers were left to travel. Let t = this number.

Length of total trip	minus	Distance traveled	is	Distance left to travel
↓	↓	↓	↓	↓
288	-	180	=	t

We carry out the subtraction:

$288 - 180 = t$

$108 = t$

Check. We repeat the calculations. The results check.

State. The total trip was 288 km. There were 108 km left to travel.

Exercise Set 3.1

In this section we will find the LCM using the multiples method in Exercises 1 - 19 and the factorization method in Exercises 21 - 43.

1. a) 4 is a multiple of 2, so it is the LCM.
   c) LCM = 4

3. a) 25 is not a multiple of 10.
   b) Check multiples:
      $2 \cdot 25 = 50$ A multiple of 10
   c) LCM = 50

5. a) 40 is a multiple of 20, so it is the LCM.
   c) LCM = 40

7. a) 27 is not a multiple of 18.
   b) Check multiples:
      $2 \cdot 27 = 54$ A multiple of 18
   c) LCM = 54

9. a) 50 is not a multiple of 30.
   b) Check multiples:
      $2 \cdot 50 = 100$ Not a multiple of 30
      $3 \cdot 50 = 150$ A multiple of 30
   c) LCM = 150

11. a) 40 is not a multiple of 30.
    b) Check multiples:
       $2 \cdot 40 = 80$ Not a multiple of 30
       $3 \cdot 40 = 120$ A multiple of 30
    c) LCM = 120

13. a) 24 is not a multiple of 18.
    b) Check multiples:
       $2 \cdot 24 = 48$ Not a multiple of 18
       $3 \cdot 24 = 72$ A multiple of 18
    c) LCM = 72

15. a) 70 is not a multiple of 60.
    b) Check multiples:
       $2 \cdot 70 = 140$ Not a multiple of 60
       $3 \cdot 70 = 210$ Not a multiple of 60
       $4 \cdot 70 = 280$ Not a multiple of 60
       $5 \cdot 70 = 350$ Not a multiple of 60
       $6 \cdot 70 = 420$ A multiple of 60
    c) LCM = 420

17. a) 36 is not a multiple of 16.
    b) Check multiples:
       $2 \cdot 36 = 72$ Not a multiple of 16
       $3 \cdot 36 = 108$ Not a multiple of 16
       $4 \cdot 36 = 144$ A multiple of 16
    c) LCM = 144

19. a) 36 is not a multiple of 32.
    b) Check multiples:
       $2 \cdot 36 = 72$ Not a multiple of 32
       $3 \cdot 36 = 108$ Not a multiple of 32
       $4 \cdot 36 = 144$ Not a multiple of 32
       $5 \cdot 36 = 180$ Not a multiple of 32
       $6 \cdot 36 = 216$ Not a multiple of 32
       $7 \cdot 36 = 252$ Not a multiple of 32
       $8 \cdot 36 = 288$ A multiple of 32
    c) LCM = 288

21. Note that each of the numbers 2, 3, and 5 is prime. They have no common prime factor. When this happens, the LCM is just the product of the numbers.

    LCM is $2 \cdot 3 \cdot 5$, or 30.

23. Note that each of the numbers 3, 5, and 7 is prime. They have no common prime factor. When this happens, the LCM is just the product of the numbers.

    LCM is $3 \cdot 5 \cdot 7$, or 105.

25. a) Find the prime factorization of each number.
       $24 = 2 \cdot 2 \cdot 2 \cdot 3$
       $36 = 2 \cdot 2 \cdot 3 \cdot 3$
       $12 = 2 \cdot 2 \cdot 3$
    b) Create a product by writing factors, using each the greatest number of times it occurs in any one factorization.

       Consider the factor 2. The greatest number of times 2 occurs in any one factorization is three. We write 2 as a factor three times.
       $2 \cdot 2 \cdot 2 \cdot$?

       Consider the factor 3. The greatest number of times 3 occurs in any one factorization is two. We write 3 as a factor two times.
       $2 \cdot 2 \cdot 2 \cdot 3 \cdot 3 \cdot$?

       Since there are no other prime factors in any of the factorizations, the LCM is $2 \cdot 2 \cdot 2 \cdot 3 \cdot 3$, or 72.

27. a) Find the prime factorization of each number.
       $5 = 5$ (5 is prime.)
       $12 = 2 \cdot 2 \cdot 3$
       $15 = 3 \cdot 5$
    b) Create a product by writing each factor the greatest number of times it occurs in any one factorization.

       The greatest number of times 2 occurs in any one factorization is two times.

       The greatest number of times 3 occurs in any one factorization is one time.

       The greatest number of times 5 occurs in any one factorization is one time.

       Since there are no other prime factors in any of the factorizations, the LCM is $2 \cdot 2 \cdot 3 \cdot 5$, or 60.

29. a) Find the prime factorization of each number.

   9 = 3·3

   12 = 2·2·3

   6 = 2·3

   b) Create a product by writing each factor the greatest number of times it occurs in any one factorization.

   The greatest number of times 2 occurs in any one factorization is two times.

   The greatest number of times 3 occurs in any one factorization is two times.

   Since there are no other prime factors in any one of the factorizations, the LCM is 2·2·3·3, or 36.

31. a) Find the prime factorization of each number.

   3 = 3   (3 is prime.)

   6 = 2·3

   8 = 2·2·2

   b) Create a product by writing each factor the greatest numbers of times it occurs in any one factorization.

   The greatest number of times 2 occurs in any one factorization is three times.

   The greatest number of times 3 occurs in any one factorization is one time.

   Since there are no other prime factor in any of the factorizations, the LCM is 2·2·2·3, or 24.

33. Note that 8 is a factor of 48. If one number is a factor of another, the LCM is the greater number.

   The LCM is 48.

   The factorization method will also work here if you do not recognize at the outset that 8 is a factor of 48.

35. Note that 5 is a factor of 50. If one number is a factor of another, the LCM is the greater number.

   The LCM is 50.

37. Note that 11 and 13 are prime. They have no common prime factor. When this happens, the LCM is just the product of the numbers.

   The LCM is 11·13, or 143.

39. a) Find the prime factorization of each number.

   12 = 2·2·3

   35 = 5·7

   b) Note that the two numbers have no common prime factor. When this happens, the LCM is just the product of the numbers.

   The LCM is 12·35, or 420.

41. a) Find the prime factorization of each number.

   54 = 3·3·3·2

   63 = 3·3·7

   b) Create a product by writing each factor the greatest number of times it occurs in any one factorization.

   The greatest number of times 2 occurs in any one factorization is one time.

   The greatest number of times 3 occurs in any one factorization is three times.

   The greatest number of times 7 occurs in any one factorization is one time.

   Since there are no other prime factors in any of the factorizations, the LCM is 2·3·3·3·7, or 378.

43. a) Find the prime factorization of each number.

   81 = 3·3·3·3

   90 = 2·3·3·5

   b) Create a product by writing each factor the greatest number of times it occurs in any one factorization.

   The greatest number of times 2 occurs in any one factorization is one time.

   The greatest number of times 3 occurs in any one factorization is four times.

   The greatest number of times 5 occurs in any one factorization is one time.

   Since there are no other prime factors in any of the factorizations, the LCM is 2·3·3·3·3·5, or 810.

45. Familiarize. We draw a picture. Repeated addition applies here.

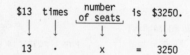

   Translate. We must determine how many 13's there are in 3250. This number is the number of seats in the auditorium. We let x = the number of seats in the auditorium.

   $13 times number of seats is $3250.

   13 · x = 3250

   Solve. To solve the equation, we divide on both sides by 13.

   $$13·x = 3250$$

   $$\frac{13·x}{13} = \frac{3250}{13}$$

   $$x = 250$$

   Check. If 250 seats are sold at $13 each, the total receipts are 250·$13, or $3250. The result checks.

   State. The auditorium contains 250 seats.

47. $\frac{4}{3} · \frac{10}{10} = \frac{4}{3} · 1 = \frac{4}{3}$

49. The length of the carton must be a multiple of both 6 and 8. The shortest length carton will be the least common multiple of 6 and 8.

$$6 = 2 \cdot 3$$
$$8 = 2 \cdot 2 \cdot 2$$

LCM is $2 \cdot 2 \cdot 2 \cdot 3$, or 24.

The shortest carton is 24 in. long.

51. a) 7800 is not a multiple of 2700.

   b) Check multiples, using a calculator:

   $2 \cdot 7800 = 15,600$ Not a multiple of 2700
   $3 \cdot 7800 = 23,400$ Not a multiple of 2700
   $4 \cdot 7800 = 31,200$ Not a multiple of 2700
   $5 \cdot 7800 = 39,000$ Not a multiple of 2700
   $6 \cdot 7800 = 46,800$ Not a multiple of 2700
   $8 \cdot 7800 = 62,400$ Not a multiple of 2700
   $9 \cdot 7800 = 70,200$ A multiple of 2700

   c) LCM is 70,200

Exercise Set 3.2

1. $\frac{7}{8} + \frac{1}{8} = \frac{7+1}{8} = \frac{8}{8} = 1$

3. $\frac{1}{8} + \frac{5}{8} = \frac{1+5}{8} = \frac{6}{8} = \frac{3 \cdot 2}{4 \cdot 2} = \frac{3}{4} \cdot \frac{2}{2} = \frac{3}{4} \cdot 1 = \frac{3}{4}$

5. $\frac{2}{3} + \frac{5}{6}$ 3 is a factor of 6, so the LCD is 6.

$= \frac{2}{3} \cdot \frac{2}{2} + \frac{5}{6}$ ⟵ This fraction already has the LCD as denominator.

Think: $3 \times \boxed{} = 6$. The answer is 2, so we multiply by 1, using $\frac{2}{2}$.

$= \frac{4}{6} + \frac{5}{6} = \frac{9}{6}$

$= \frac{3}{2}$ Simplifying

7. $\frac{1}{8} + \frac{1}{6}$ $8 = 2 \cdot 2 \cdot 2$ and $6 = 2 \cdot 3$, so the LCD is $2 \cdot 2 \cdot 2 \cdot 3$, or 24.

$= \frac{1}{8} \cdot \frac{3}{3} + \frac{1}{6} \cdot \frac{4}{4}$

Think: $6 \times \boxed{} = 24$. The answer is 4, so we multiply by 1, using $\frac{4}{4}$.

Think: $8 \times \boxed{} = 24$. The answer is 3, so we multiply by 1, using $\frac{3}{3}$.

$= \frac{3}{24} + \frac{4}{24}$

$= \frac{7}{24}$

9. $\frac{4}{5} + \frac{7}{10}$ 5 is a factor of 10, so the LCD is 10.

$= \frac{4}{5} \cdot \frac{2}{2} + \frac{7}{10}$ ⟵ This fraction already has the LCD as denominator.

Think: $5 \times \boxed{} = 10$. The answer is 2, so we multiply by 1, using $\frac{2}{2}$.

$= \frac{8}{10} + \frac{7}{10} = \frac{15}{10}$

$= \frac{3}{2}$ Simplifying

11. $\frac{5}{12} + \frac{3}{8}$ $12 = 2 \cdot 2 \cdot 3$ and $8 = 2 \cdot 2 \cdot 2$, so the LCD is $2 \cdot 2 \cdot 2 \cdot 3$, or 24

$= \frac{5}{12} \cdot \frac{2}{2} + \frac{3}{8} \cdot \frac{3}{3}$

Think: $8 \times \boxed{} = 24$. The answer is 3, so we multiply by 1, using $\frac{3}{3}$.

Think: $12 \times \boxed{} = 24$. The answer is 2, so we multiply by 1, using $\frac{2}{2}$.

$= \frac{10}{24} + \frac{9}{24} = \frac{19}{24}$

13. $\frac{3}{20} + \frac{3}{4}$ 4 is a factor of 20, so the LCD is 20.

$= \frac{3}{20} + \frac{3}{4} \cdot \frac{5}{5}$ Multiplying by 1

$= \frac{3}{20} + \frac{15}{20} = \frac{18}{20} = \frac{9}{10}$

15. $\frac{5}{6} + \frac{7}{9}$ $6 = 2 \cdot 3$ and $9 = 3 \cdot 3$, so the LCD is $2 \cdot 3 \cdot 3$, or 18

$= \frac{5}{6} \cdot \frac{3}{3} + \frac{7}{9} \cdot \frac{2}{2}$ Multiplying by 1

$= \frac{15}{18} + \frac{14}{18} = \frac{29}{18}$

17. $\frac{3}{10} + \frac{1}{100}$ 10 is a factor of 100, so the LCD is 100.

$= \frac{3}{10} \cdot \frac{10}{10} + \frac{1}{100}$

$= \frac{30}{100} + \frac{1}{100} = \frac{31}{100}$

19. $\frac{5}{12} + \frac{4}{15}$ $12 = 2 \cdot 2 \cdot 3$ and $15 = 3 \cdot 5$, so the LCD is $2 \cdot 2 \cdot 3 \cdot 5$, or 60.

$= \frac{5}{12} \cdot \frac{5}{5} + \frac{4}{15} \cdot \frac{4}{4}$

$= \frac{25}{60} + \frac{16}{60} = \frac{41}{60}$

21. $\frac{9}{10} + \frac{99}{100}$ 10 is a factor of 100, so the LCD is 100.

$= \frac{9}{10} \cdot \frac{10}{10} + \frac{99}{100}$

$= \frac{90}{100} + \frac{99}{100} = \frac{189}{100}$

23. $\frac{7}{8} + \frac{0}{1}$ 1 is a factor of 8, so the LCD is 8.

$= \frac{7}{8} + \frac{0}{1} \cdot \frac{8}{8}$

$= \frac{7}{8} + \frac{0}{8} = \frac{7}{8}$

Note that if we had observed at the outset that $\frac{0}{1} = 0$, the computation becomes $\frac{7}{8} + 0 = \frac{7}{8}$.

25. $\frac{3}{8} + \frac{1}{6}$ $8 = 2 \cdot 2 \cdot 2$ and $6 = 2 \cdot 3$, so the LCD is $2 \cdot 2 \cdot 2 \cdot 3$, or 24.

$= \frac{3}{8} \cdot \frac{3}{3} + \frac{1}{6} \cdot \frac{4}{4}$

$= \frac{9}{24} + \frac{4}{24} = \frac{13}{24}$

27. $\frac{5}{12} + \frac{7}{24}$ 12 is a factor of 24, so the LCD is 24.

$= \frac{5}{12} \cdot \frac{2}{2} + \frac{7}{24}$

$= \frac{10}{24} + \frac{7}{24} = \frac{17}{24}$

29. $\frac{3}{16} + \frac{5}{16} + \frac{4}{16} = \frac{3 + 5 + 4}{16} = \frac{12}{16} = \frac{3}{4}$

31. $\frac{8}{10} + \frac{7}{100} + \frac{4}{1000}$ 10 and 100 are factors of 1000, so the LCD is 1000.

$= \frac{8}{10} \cdot \frac{100}{100} + \frac{7}{100} \cdot \frac{10}{10} + \frac{4}{1000}$

$= \frac{800}{1000} + \frac{70}{1000} + \frac{4}{1000} = \frac{874}{1000}$

$= \frac{437}{500}$

33. $\frac{3}{8} + \frac{5}{12} + \frac{8}{15}$

$= \frac{3}{2 \cdot 2 \cdot 2} + \frac{5}{2 \cdot 2 \cdot 3} + \frac{8}{3 \cdot 5}$ Factoring the denominators

The LCM is $2 \cdot 2 \cdot 2 \cdot 3 \cdot 5$, or 120.

$= \frac{3}{2 \cdot 2 \cdot 2} \cdot \frac{3 \cdot 5}{3 \cdot 5} + \frac{5}{2 \cdot 2 \cdot 3} \cdot \frac{2 \cdot 5}{2 \cdot 5} + \frac{8}{3 \cdot 5} \cdot \frac{2 \cdot 2 \cdot 2}{2 \cdot 2 \cdot 2}$

In each case we multiply by 1 to obtain the LCD in the denominator.

33. (continued)

$= \frac{3 \cdot 3 \cdot 5}{2 \cdot 2 \cdot 2 \cdot 3 \cdot 5} + \frac{5 \cdot 2 \cdot 5}{2 \cdot 2 \cdot 3 \cdot 2 \cdot 5} + \frac{8 \cdot 2 \cdot 2 \cdot 2}{3 \cdot 5 \cdot 2 \cdot 2 \cdot 2}$

$= \frac{45}{120} + \frac{50}{120} + \frac{64}{120}$

$= \frac{159}{120} = \frac{53}{40}$

35. $\frac{15}{24} + \frac{7}{36} + \frac{91}{48}$

$= \frac{15}{2 \cdot 2 \cdot 2 \cdot 3} + \frac{7}{2 \cdot 2 \cdot 3 \cdot 3} + \frac{91}{2 \cdot 2 \cdot 2 \cdot 2 \cdot 3}$ Factoring the denominators

The LCM is $2 \cdot 2 \cdot 2 \cdot 2 \cdot 3 \cdot 3$, or 144.

$= \frac{15}{2 \cdot 2 \cdot 2 \cdot 3} \cdot \frac{2 \cdot 3}{2 \cdot 3} + \frac{7}{2 \cdot 2 \cdot 3 \cdot 3} \cdot \frac{2 \cdot 2}{2 \cdot 2} + \frac{91}{2 \cdot 2 \cdot 2 \cdot 2 \cdot 3} \cdot \frac{3}{3}$

In each case we multiply by 1 to get the LCD in the denominator.

$= \frac{15 \cdot 2 \cdot 3}{2 \cdot 2 \cdot 2 \cdot 3 \cdot 2 \cdot 3} + \frac{7 \cdot 2 \cdot 2}{2 \cdot 2 \cdot 3 \cdot 3 \cdot 2 \cdot 2} + \frac{91 \cdot 3}{2 \cdot 2 \cdot 2 \cdot 2 \cdot 3 \cdot 3}$

$= \frac{90}{144} + \frac{28}{144} + \frac{273}{144} = \frac{391}{144}$

37. Familiarize. We draw a picture. We let p = the number of pounds of candy the consumer bought.

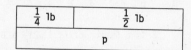

$\frac{1}{4}$ lb	$\frac{1}{2}$ lb
	p

Translate. An addition sentence corresponds to this situation.

Pounds of bonbons	plus	Pounds of caramels	is	Total pounds of candy
↓	↓	↓	↓	↓
$\frac{1}{4}$	+	$\frac{1}{2}$	=	p

Solve. To solve the equation we carry out the addition. The LCM of the denominators is 4 since 2 is a factor of 4.

$\frac{1}{4} + \frac{1}{2} \cdot \frac{2}{2} = p$

$\frac{1}{4} + \frac{2}{4} = p$

$\frac{3}{4} = p$

Check. We check by repeating the calculation. We also note that the sum is larger than either of the individual weights, so the answer seems reasonable.

State. The consumer bought $\frac{3}{4}$ lb of candy.

39. <u>Familiarize</u>. We draw a picture. We let D = the total distance walked.

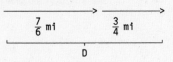

<u>Translate</u>. An addition sentence corresponds to this situation.

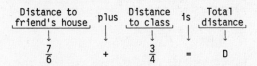

<u>Solve</u>. To solve the equation, carry out the addition. Since 6 = 2·3 and 4 = 2·2, the LCM of the denominators is 2·2·3, or 12.

$$\frac{7}{6} \cdot \frac{2}{2} + \frac{3}{4} \cdot \frac{3}{3} = D$$

$$\frac{14}{12} + \frac{9}{12} = D$$

$$\frac{23}{12} = D$$

<u>Check</u>. We repeat the calculation. We also note that the sum is larger than either of the original distances, so the answer seems reasonable.

<u>State</u>. The student walked $\frac{23}{12}$ mi.

41. First we find the amount of liquid needed for the recipe.

<u>Familiarize</u>. We draw a picture. We let a = the total amount of liquid needed.

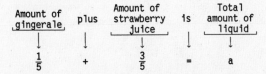

<u>Translate</u>. An addition sentence corresponds to this situation.

Amount of gingerale	plus	Amount of strawberry juice	is	Total amount of liquid
↓	↓	↓	↓	↓
$\frac{1}{5}$	+	$\frac{3}{5}$	=	a

<u>Solve</u>. To solve the equation, carry out the addition. Since the denominators are the same, we add the numerators and keep the denominator.

$$\frac{3+1}{5} = a$$

$$\frac{4}{5} = a$$

<u>Check</u>. We repeat the calculation. We also note that the sum is larger than either of the individual amounts, so the answer seems reasonable.

41. (continued)

<u>State</u>. $\frac{4}{5}$ L of liquid was needed.

Next we find the amount of liquid needed if the recipe is doubled.

<u>Familiarize</u>. We let x = the amount of liquid needed if the recipe is doubled.

<u>Translate</u>. A multiplication sentence applies.

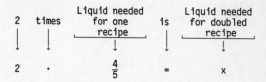

<u>Solve</u>. To solve the equation, carry out the multiplication.

$$\frac{2 \cdot 4}{5} = x$$

$$\frac{8}{5} = x$$

<u>Check</u>. We repeat the calculation. We also note that the new amount is larger than the original amount, so the answer seems reasonable.

<u>State</u>. If the recipe is doubled, $\frac{8}{5}$ L of liquid is needed.

Finally, we find the amount of liquid needed if the recipe is halved.

<u>Familiarize</u>. Let y = the amount of liquid needed if the recipe is halved.

<u>Translate</u>. Another multiplication sentence applies.

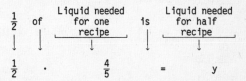

<u>Solve</u>. To solve the equation, carry out the multiplication.

$$\frac{1 \cdot 4}{2 \cdot 5} = y$$

$$\frac{1 \cdot 2 \cdot 2}{2 \cdot 5} = y$$

$$\frac{2}{2} \cdot \frac{1 \cdot 2}{5} = y$$

$$\frac{1 \cdot 2}{5} = y$$

$$\frac{2}{5} = y$$

<u>Check</u>. We repeat the calculation. We also note that the new amount is less than the original amount, so the answer seems reasonable.

<u>State</u>. If the recipe is halved, $\frac{2}{5}$ L of liquid is needed.

43. _Familiarize_. We draw a picture. We let t = the total thickness.

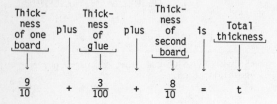

Translate. An addition sentence corresponds to this situation.

Thick-ness of one board	plus	Thick-ness of glue	plus	Thick-ness of second board	is	Total thickness
↓	↓	↓	↓	↓	↓	↓
$\frac{9}{10}$	+	$\frac{3}{100}$	+	$\frac{8}{10}$	=	t

Solve. To solve the equation, carry out the addition.

The LCD is 100 since 10 is a factor of 100.

$$\frac{9}{10} \cdot \frac{10}{10} + \frac{3}{100} + \frac{8}{10} \cdot \frac{10}{10} = t$$

$$\frac{90}{100} + \frac{3}{100} + \frac{80}{100} = t$$

$$\frac{173}{100} = t$$

Check. We repeat the calculation. We also note that the sum is larger than any of the individual thicknesses, so the answer seems reasonable.

State. The result is $\frac{173}{100}$ cm thick.

45. _Familiarize_. We let t = the part of the job that is done in total on the two days.
Translate.

One-half	of	four-ninths	plus	two-fifths	of	five-ninths	is	part of job done on the two days
↓		↓		↓		↓		↓
$\frac{1}{2}$	⋅	$\frac{4}{9}$	+	$\frac{2}{5}$	⋅	$\frac{5}{9}$	=	t

Solve. We solve the equation.

$$\frac{1}{2} \cdot \frac{4}{9} + \frac{2}{5} \cdot \frac{5}{9} = t$$

$$\frac{1 \cdot 4}{2 \cdot 9} + \frac{2 \cdot 5}{5 \cdot 9} = t \quad \text{Multiplying}$$

$$\frac{1 \cdot 2 \cdot 2}{2 \cdot 9} + \frac{2 \cdot 5}{5 \cdot 9} = t$$

$$\frac{1 \cdot 2}{9} \cdot \frac{2}{2} + \frac{2}{9} \cdot \frac{5}{5} = t$$

$$\frac{2}{9} + \frac{2}{9} = t \quad \text{Simplifying}$$

$$\frac{4}{9} = t \quad \text{Adding}$$

Check. We repeat the calculations.

45. (continued)

State. $\frac{4}{9}$ of the job is done in total on the two days.

Exercise Set 3.3

1. When denominators are the same, subtract the numerators and keep the denominator.

$$\frac{5}{6} - \frac{1}{6} = \frac{5-1}{6} = \frac{4}{6} = \frac{2}{3}$$

3. When denominators are the same, subtract the numerators and keep the denominator.

$$\frac{11}{12} - \frac{2}{12} = \frac{11-2}{12} = \frac{9}{12} = \frac{3}{4}$$

5. The LCM of 4 and 8 is 8.

$$\frac{3}{4} - \frac{1}{8} = \frac{3}{4} \cdot \frac{2}{2} - \frac{1}{8} \leftarrow \text{This fraction already has the LCM as the denominator.}$$

Think: $4 \times \boxed{} = 8$. The answer is 2, so we multiply by 1, using $\frac{2}{2}$.

$$= \frac{6}{8} - \frac{1}{8} = \frac{5}{8}$$

7. The LCM of 8 and 12 is 24.

$$\frac{1}{8} - \frac{1}{12} = \frac{1}{8} \cdot \frac{3}{3} - \frac{1}{12} \cdot \frac{2}{2}$$

Think: $12 \times \boxed{} = 24$. The answer is 2, so we multiply by 1, using $\frac{2}{2}$.

Think: $8 \times \boxed{} = 24$. The answer is 3, so we multiply by 1, using $\frac{3}{3}$.

$$= \frac{3}{24} - \frac{2}{24} = \frac{1}{24}$$

9. The LCM of 3 and 6 is 6.

$$\frac{4}{3} - \frac{5}{6} = \frac{4}{3} \cdot \frac{2}{2} - \frac{5}{6}$$

$$= \frac{8}{6} - \frac{5}{6} = \frac{3}{6}$$

$$= \frac{1}{2}$$

11. The LCM of 4 and 28 is 28.

$$\frac{3}{4} - \frac{3}{28} = \frac{3}{4} \cdot \frac{7}{7} - \frac{3}{28}$$

$$= \frac{21}{28} - \frac{3}{28}$$

$$= \frac{18}{28} = \frac{9 \cdot 2}{14 \cdot 2}$$

$$= \frac{9}{14} \cdot \frac{2}{2} = \frac{9}{14} \cdot 1$$

$$= \frac{9}{14}$$

13. The LCM of 4 and 20 is 20.

$$\frac{3}{4} - \frac{3}{20} = \frac{3}{4} \cdot \frac{5}{5} - \frac{3}{20}$$

$$= \frac{15}{20} - \frac{3}{20} = \frac{12}{20}$$

$$= \frac{3}{5}$$

15. The LCM of 4 and 20 is 20.

$$\frac{3}{4} - \frac{1}{20} = \frac{3}{4} \cdot \frac{5}{5} - \frac{1}{20}$$

$$= \frac{15}{20} - \frac{1}{20}$$

$$= \frac{14}{20} = \frac{2 \cdot 7}{2 \cdot 10}$$

$$= \frac{7}{10}$$

17. The LCM of 12 and 15 is 60.

$$\frac{5}{12} - \frac{2}{15} = \frac{5}{12} \cdot \frac{5}{5} - \frac{2}{15} \cdot \frac{4}{4}$$

$$= \frac{25}{60} - \frac{8}{60} = \frac{17}{60}$$

19. The LCM of 10 and 100 is 100.

$$\frac{6}{10} - \frac{7}{100} = \frac{6}{10} \cdot \frac{10}{10} - \frac{7}{100}$$

$$= \frac{60}{100} - \frac{7}{100} = \frac{53}{100}$$

21. The LCM of 15 and 25 is 75.

$$\frac{7}{15} - \frac{3}{25} = \frac{7}{15} \cdot \frac{5}{5} - \frac{3}{25} \cdot \frac{3}{3}$$

$$= \frac{35}{75} - \frac{9}{75} = \frac{26}{75}$$

23. The LCM of 10 and 100 is 100.

$$\frac{99}{100} - \frac{9}{10} = \frac{99}{100} - \frac{9}{10} \cdot \frac{10}{10}$$

$$= \frac{99}{100} - \frac{90}{100} = \frac{9}{100}$$

25. The LCM of 3 and 8 is 24.

$$\frac{2}{3} - \frac{1}{8} = \frac{2}{3} \cdot \frac{8}{8} - \frac{1}{8} \cdot \frac{3}{3}$$

$$= \frac{16}{24} - \frac{3}{24}$$

$$= \frac{13}{24}$$

27. The LCM of 5 and 2 is 10.

$$\frac{3}{5} - \frac{1}{2} = \frac{3}{5} \cdot \frac{2}{2} - \frac{1}{2} \cdot \frac{5}{5}$$

$$= \frac{6}{10} - \frac{5}{10}$$

$$= \frac{1}{10}$$

29. The LCM of 12 and 8 is 24.

$$\frac{5}{12} - \frac{3}{8} = \frac{5}{12} \cdot \frac{2}{2} - \frac{3}{8} \cdot \frac{3}{3}$$

$$= \frac{10}{24} - \frac{9}{24}$$

$$= \frac{1}{24}$$

31. The LCM of 8 and 16 is 16.

$$\frac{7}{8} - \frac{1}{16} = \frac{7}{8} \cdot \frac{2}{2} - \frac{1}{16}$$

$$= \frac{14}{16} - \frac{1}{16}$$

$$= \frac{13}{16}$$

33. The LCM of 25 and 15 is 75.

$$\frac{17}{25} - \frac{4}{15} = \frac{17}{25} \cdot \frac{3}{3} - \frac{4}{15} \cdot \frac{5}{5}$$

$$= \frac{51}{75} - \frac{20}{75}$$

$$= \frac{31}{75}$$

35. The LCM of 25 and 150 is 150.

$$\frac{23}{25} - \frac{112}{150} = \frac{23}{25} \cdot \frac{6}{6} - \frac{112}{150}$$

$$= \frac{138}{150} - \frac{112}{150}$$

$$= \frac{26}{150} = \frac{2 \cdot 13}{2 \cdot 75}$$

$$= \frac{13}{75}$$

37. Since there is a common denominator, compare the numerators.

$$5 < 6, \text{ so } \frac{5}{8} < \frac{6}{8}.$$

39. The LCD is 12.

$$\frac{1}{3} \cdot \frac{4}{4} = \frac{4}{12}$$ We multiply by 1 to make the denominators the same.

$$\frac{1}{4} \cdot \frac{3}{3} = \frac{3}{12}$$

Since $4 > 3$, it follows that $\frac{4}{12} > \frac{3}{12}$, so $\frac{1}{3} > \frac{1}{4}$.

41. The LCD is 21.

$$\frac{2}{3} \cdot \frac{7}{7} = \frac{14}{21}$$ We multiply by 1 to make the denominators the same.

$$\frac{5}{7} \cdot \frac{3}{3} = \frac{15}{21}$$

Since $14 < 15$, it follows that $\frac{14}{21} < \frac{15}{21}$, so $\frac{2}{3} < \frac{5}{7}$.

43. The LCD is 30.

$$\frac{4}{5} \cdot \frac{6}{6} = \frac{24}{30}$$

$$\frac{5}{6} \cdot \frac{5}{5} = \frac{25}{30}$$

Since $24 < 25$, it follows that $\frac{24}{30} < \frac{25}{30}$, so $\frac{4}{5} < \frac{5}{6}$.

45. The LCD is 20.

The denominator of $\frac{19}{20}$ is the LCD.

$$\frac{4}{5} \cdot \frac{4}{4} = \frac{16}{20}$$

Since $19 > 16$, it follows that $\frac{19}{20} > \frac{16}{20}$, so $\frac{19}{20} > \frac{4}{5}$.

47. The LCD is 20.

The denominator of $\frac{19}{20}$ is the LCD.

$$\frac{9}{10} \cdot \frac{2}{2} = \frac{18}{20}$$

Since $19 > 18$, it follows that $\frac{19}{20} > \frac{18}{20}$, so $\frac{19}{20} > \frac{9}{10}$.

49. The LCD is $21 \cdot 13$, or 273.

$$\frac{31}{21} \cdot \frac{13}{13} = \frac{403}{273}$$

$$\frac{41}{13} \cdot \frac{21}{21} = \frac{861}{273}$$

Since $403 < 861$, it follows that $\frac{403}{273} < \frac{861}{273}$, so $\frac{31}{21} < \frac{41}{13}$.

51.
$$x + \frac{1}{30} = \frac{1}{10}$$

$$x + \frac{1}{30} - \frac{1}{30} = \frac{1}{10} - \frac{1}{30}$$ Subtracting $\frac{1}{30}$ on both sides

$$x + 0 = \frac{1}{10} \cdot \frac{3}{3} - \frac{1}{30}$$ The LCD is 30. We multiply by 1 to get the LCD.

$$x = \frac{3}{30} - \frac{1}{30}$$

$$x = \frac{2}{30} = \frac{1}{15}$$

The solution is $\frac{1}{15}$.

53.
$$\frac{2}{3} + t = \frac{4}{5}$$

$$\frac{2}{3} + t - \frac{2}{3} = \frac{4}{5} - \frac{2}{3}$$ Subtracting $\frac{2}{3}$ on both sides

$$t + 0 = \frac{4}{5} \cdot \frac{3}{3} - \frac{2}{3} \cdot \frac{5}{5}$$ The LCD is 15. We multiply by 1 to get the LCD.

$$t = \frac{12}{15} - \frac{10}{15} = \frac{2}{15}$$

The solution is $\frac{2}{15}$.

55.
$$m + \frac{5}{6} = \frac{9}{10}$$

$$m + \frac{5}{6} - \frac{5}{6} = \frac{9}{10} - \frac{5}{6}$$

$$m + 0 = \frac{9}{10} \cdot \frac{3}{3} - \frac{5}{6} \cdot \frac{5}{5}$$

$$m = \frac{27}{30} - \frac{25}{30}$$

$$m = \frac{2}{30} = \frac{1}{15}$$

The solution is $\frac{1}{15}$.

57. <u>Familiarize</u>. We draw a picture. We let x = the portion of the business owned by the third person.

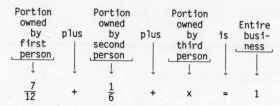

<u>Translate</u>. We have a "how much more" situation. We translate to an equation.

Portion owned by first person	plus	Portion owned by second person	plus	Portion owned by third person	is	Entire business
$\frac{7}{12}$	+	$\frac{1}{6}$	+	x	=	1

<u>Solve</u>. To solve the equation, we subtract $\frac{7}{12}$ and $\frac{1}{6}$ on both sides.

$$\frac{7}{12} + \frac{1}{6} + x = 1$$

$$\frac{7}{12} + \frac{1}{6} + x - \frac{7}{12} - \frac{1}{6} = 1 - \frac{7}{12} - \frac{1}{6}$$

$$x + 0 = 1 \cdot \frac{12}{12} - \frac{7}{12} - \frac{1}{6} \cdot \frac{2}{2}$$

The LCD is 12.

$$x = \frac{12}{12} - \frac{7}{12} - \frac{2}{12}$$

$$x = \frac{3}{12} = \frac{1}{4}$$

<u>Check</u>. We return to the original problem and add:
$$\frac{7}{12} + \frac{1}{6} + \frac{1}{4} = \frac{7}{12} + \frac{1}{6} \cdot \frac{2}{2} + \frac{1}{4} \cdot \frac{3}{3} = \frac{7}{12} + \frac{2}{12} + \frac{3}{12} = \frac{12}{12} = 1.$$ This checks.

<u>State</u>. The third person owned $\frac{1}{4}$ of the business.

59. $\frac{9}{10} \div \frac{3}{5} = \frac{9}{10} \cdot \frac{5}{3} = \frac{9 \cdot 5}{10 \cdot 3} = \frac{3 \cdot 3 \cdot 5}{2 \cdot 5 \cdot 3} = \frac{3 \cdot 5}{3 \cdot 5} \cdot \frac{3}{2} = \frac{3}{2}$

61. The batting average is a fraction with the number of hits as its numerator and the number of at bats as its denominator.

Jim Rice's batting average was $\frac{174}{564}$.

George Brett's batting average was $\frac{195}{634}$.

Using a calculator, we multiply by one to make the denominators the same.

$$\frac{174}{564} \cdot \frac{634}{634} = \frac{110,316}{357,576}$$

$$\frac{195}{634} \cdot \frac{564}{564} = \frac{109,980}{357,576}$$

Since $110,316 > 109,980$, it follows that $\frac{110,316}{357,576} > \frac{109,980}{357,576}$, so $\frac{174}{564} > \frac{195}{634}$. Thus, Jim Rice had the higher batting average.

63. $\frac{7}{8} - \frac{1}{10} \times \frac{5}{6} = \frac{7}{8} - \frac{1 \times 5}{10 \times 6} = \frac{7}{8} - \frac{1 \times 5}{2 \times 5 \times 6}$

$= \frac{7}{8} - \frac{5}{5} \times \frac{1}{2 \times 6} = \frac{7}{8} - \frac{1}{2 \times 6} = \frac{7}{8} - \frac{1}{12}$

$= \frac{7}{8} \cdot \frac{3}{3} - \frac{1}{12} \cdot \frac{2}{2} = \frac{21}{24} - \frac{2}{24} = \frac{19}{24}$

65. $\left(\frac{2}{3}\right)^2 + \left(\frac{3}{4}\right)^2 = \frac{4}{9} + \frac{9}{16} = \frac{4}{9} \cdot \frac{16}{16} + \frac{9}{16} \cdot \frac{9}{9} =$

$\frac{64}{144} + \frac{81}{144} = \frac{145}{144}$

Exercise Set 3.4

1. $5\frac{2}{3} = \frac{17}{3}$ ① Multiply: $5 \cdot 3 = 15$

   ② Add: $15 + 2 = 17$

   ③ Keep the denominator

3. $6\frac{1}{4} = \frac{25}{4}$ ① Multiply: $6 \cdot 4 = 24$

   ② Add: $24 + 1 = 25$

   ③ Keep the denominator

5. $10\frac{1}{8} = \frac{81}{8}$ $(10 \cdot 8 = 80, \; 80 + 1 = 81)$

7. $5\frac{1}{10} = \frac{51}{10}$ $(5 \cdot 10 = 50, \; 50 + 1 = 51)$

9. $20\frac{3}{5} = \frac{103}{5}$ $(20 \cdot 5 = 100, \; 100 + 3 = 103)$

11. $9\frac{5}{6} = \frac{59}{6}$ $(9 \cdot 6 = 54, \; 54 + 5 = 59)$

13. $7\frac{3}{10} = \frac{73}{10}$ $(7 \cdot 10 = 70, \; 70 + 3 = 73)$

15. $1\frac{5}{8} = \frac{13}{8}$ $(1 \cdot 8 = 8, \; 8 + 5 = 13)$

17. $12\frac{3}{4} = \frac{51}{4}$ $(12 \cdot 4 = 48, \; 48 + 3 = 51)$

19. $4\frac{3}{10} = \frac{43}{10}$ $(4 \cdot 10 = 40, \; 40 + 3 = 43)$

21. $2\frac{3}{100} = \frac{203}{100}$ $(2 \cdot 100 = 200, \; 200 + 3 = 203)$

23. $66\frac{2}{3} = \frac{200}{3}$ $(66 \cdot 3 = 198, \; 198 + 2 = 200)$

25. $5\frac{29}{50} = \frac{279}{50}$ $(5 \cdot 50 = 250, \; 250 + 29 = 279)$

27. To convert $\frac{8}{5}$ to a mixed numeral, we divide.

$$
\begin{array}{r}
1 \\
5 \overline{)8} \\
5 \\
\hline
3
\end{array}
\qquad \frac{8}{5} = 1\frac{3}{5}
$$

29. To convert $\frac{14}{3}$ to a mixed numeral, we divide.

$$
\begin{array}{r}
4 \\
3 \overline{)14} \\
12 \\
\hline
2
\end{array}
\qquad \frac{14}{3} = 4\frac{2}{3}
$$

31. $$
\begin{array}{r}
4 \\
6 \overline{)27} \\
24 \\
\hline
3
\end{array}
\qquad \frac{27}{6} = 4\frac{3}{6} = 4\frac{1}{2}
$$

33. $$
\begin{array}{r}
5 \\
10 \overline{)57} \\
50 \\
\hline
7
\end{array}
\qquad \frac{57}{10} = 5\frac{7}{10}
$$

35. $$
\begin{array}{r}
7 \\
7 \overline{)53} \\
49 \\
\hline
4
\end{array}
\qquad \frac{53}{7} = 7\frac{4}{7}
$$

37. $$
\begin{array}{r}
7 \\
6 \overline{)45} \\
42 \\
\hline
3
\end{array}
\qquad \frac{45}{6} = 7\frac{3}{6} = 7\frac{1}{2}
$$

39. $$
\begin{array}{r}
11 \\
4 \overline{)46} \\
40 \\
\hline
6 \\
4 \\
\hline
2
\end{array}
\qquad \frac{46}{4} = 11\frac{2}{4} = 11\frac{1}{2}
$$

41. $$
\begin{array}{r}
1 \\
8 \overline{)12} \\
8 \\
\hline
4
\end{array}
\qquad \frac{12}{8} = 1\frac{4}{8} = 1\frac{1}{2}
$$

43.

$$100\overline{)757} \quad \begin{array}{r} 7 \\ \hline 7\,5\,7 \\ 7\,0\,0 \\ \hline 5\,7 \end{array} \qquad \frac{757}{100} = 7\frac{57}{100}$$

45.

$$8\overline{)345} \quad \begin{array}{r} 4\,3 \\ \hline 3\,4\,5 \\ 3\,2\,0 \\ \hline 2\,5 \\ 2\,4 \\ \hline 1 \end{array} \qquad \frac{345}{8} = 43\frac{1}{8}$$

47. We first divide as usual.

$$8\overline{)869} \quad \begin{array}{r} 1\,0\,8 \\ \hline 8\,6\,9 \\ 8\,0\,0 \\ \hline 6\,9 \\ 6\,4 \\ \hline 5 \end{array}$$

The answer is 108 R 5. We write a mixed numeral for the quotient as follows: $108\frac{5}{8}$

49. We first divide as usual.

$$7\overline{)6345} \quad \begin{array}{r} 9\,0\,6 \\ \hline 6\,3\cdot4\,5 \\ 6\,3\,0\,0 \\ \hline 4\,5 \\ 4\,2 \\ \hline 3 \end{array}$$

The answer is 906 R 3. We write a mixed numeral for the quotient as follows: $906\frac{3}{7}$

51.

$$21\overline{)852} \quad \begin{array}{r} 4\,0 \\ \hline 8\,5\,2 \\ 8\,4\,0 \\ \hline 1\,2 \end{array}$$

We get $40\frac{12}{21}$. This simplifies as $40\frac{4}{7}$.

53.

$$102\overline{)5612} \quad \begin{array}{r} 5\,5 \\ \hline 5\,6\,1\,2 \\ 5\,1\,0\,0 \\ \hline 5\,1\,2 \\ 5\,1\,0 \\ \hline 2 \end{array}$$

We get $55\frac{2}{102}$. This simplifies as $55\frac{1}{51}$.

55. $\frac{6}{5} \cdot 15 = \frac{6 \cdot 15}{5} = \frac{6 \cdot 3 \cdot 5}{5 \cdot 1} = \frac{5}{5} \cdot \frac{6 \cdot 3}{1} = \frac{6 \cdot 3}{1} = \frac{18}{1} = 18$

57. $\frac{7}{10} \cdot \frac{5}{14} = \frac{7 \cdot 5}{10 \cdot 14} = \frac{7 \cdot 5 \cdot 1}{2 \cdot 5 \cdot 2 \cdot 7} = \frac{7 \cdot 5}{7 \cdot 5} \cdot \frac{1}{2 \cdot 2} = \frac{1}{2 \cdot 2} = \frac{1}{4}$

59. $\frac{56}{7} + \frac{2}{3} = 8 + \frac{2}{3} \qquad (56 \div 7 = 8)$

$\qquad = 8\frac{2}{3}$

61.

$$7\overline{)366} \quad \begin{array}{r} 5\,2 \\ \hline 3\,6\,6 \\ 3\,5\,0 \\ \hline 1\,6 \\ 1\,4 \\ \hline 2 \end{array} \qquad \frac{366}{7} = 52\frac{2}{7}$$

Exercise Set 3.5

1. $\begin{array}{r} 2\frac{7}{8} \\ + 3\frac{5}{8} \\ \hline 5\frac{12}{8} \end{array} = 5 + \frac{12}{8}$

To find a mixed numeral for $\frac{12}{8}$ we divide:

$$8\overline{)12} \quad \begin{array}{r} 1 \\ \hline 1\,2 \\ 8 \\ \hline 4 \end{array} \qquad \frac{12}{8} = 1\frac{4}{8} = 1\frac{1}{2}$$

$\qquad = 5 + 1\frac{1}{2}$

$\qquad = 6\frac{1}{2}$

3. The LCD is 12.

$\begin{array}{r} 1\,\boxed{\frac{1}{4} \cdot \frac{3}{3}} = 1\,\frac{3}{12} \\ + 1\,\boxed{\frac{2}{3} \cdot \frac{4}{4}} = + 1\,\frac{8}{12} \\ \hline 2\,\frac{11}{12} \end{array}$

5. The LCD is 12.

$\begin{array}{r} 8\,\boxed{\frac{3}{4} \cdot \frac{3}{3}} = 8\,\frac{9}{12} \\ + 5\,\boxed{\frac{5}{6} \cdot \frac{2}{2}} = + 5\,\frac{10}{12} \\ \hline 13\,\frac{19}{12} \end{array} = 13 + \frac{19}{12}$

$\qquad\qquad = 13 + 1\frac{7}{12}$

$\qquad\qquad = 14\frac{7}{12}$

7. The LCD is 10.

$\begin{array}{r} 3\,\boxed{\frac{2}{5} \cdot \frac{2}{2}} = 3\,\frac{4}{10} \\ + 8\,\frac{7}{10} = + 8\,\frac{7}{10} \\ \hline 11\,\frac{11}{10} \end{array} = 11 + \frac{11}{10}$

$\qquad\qquad = 11 + 1\frac{1}{10}$

$\qquad\qquad = 12\frac{1}{10}$

9. $\begin{array}{r} 5\,\boxed{\frac{3}{8} \cdot \frac{3}{3}} = 5\,\frac{9}{24} \\ + 10\,\boxed{\frac{5}{6} \cdot \frac{4}{4}} = + 10\,\frac{20}{24} \\ \hline 15\,\frac{29}{24} \end{array} = 15 + \frac{29}{24}$

$\qquad\qquad = 15 + 1\frac{5}{24}$

$\qquad\qquad = 16\frac{5}{24}$

11. $12 \boxed{\dfrac{4}{5} \cdot \dfrac{2}{2}} = 12 \dfrac{8}{10}$

$+ \; 8 \boxed{\dfrac{7}{10}} = + \; 8 \dfrac{7}{10}$

$20 \dfrac{15}{10} = 20 + \dfrac{15}{10}$

$= 20 + 1\dfrac{5}{10}$

$= 21\dfrac{5}{10}$

$= 21\dfrac{1}{2}$

13. The LCD is 8.

$14 \dfrac{5}{8} \qquad = \; 14 \dfrac{5}{8}$

$+ \; 13 \boxed{\dfrac{1}{4} \cdot \dfrac{2}{2}} = + \; 13 \dfrac{2}{8}$

$27 \dfrac{7}{8}$

15. The LCD is 24.

$7 \boxed{\dfrac{1}{8} \cdot \dfrac{3}{3}} = \; 7 \dfrac{3}{24}$

$9 \boxed{\dfrac{2}{3} \cdot \dfrac{8}{8}} = \; 9 \dfrac{16}{24}$

$+ \; 10 \boxed{\dfrac{3}{4} \cdot \dfrac{6}{6}} = + \; 10 \dfrac{18}{24}$

$26 \dfrac{37}{24} = 26 + \dfrac{37}{24}$

$= 26 + 1\dfrac{13}{24}$

$= 27\dfrac{13}{24}$

17. $4 \dfrac{1}{5} = \; 3 \dfrac{6}{5}$

$- \; 2 \dfrac{3}{5} = - \; 2 \dfrac{3}{5}$

$1 \dfrac{3}{5}$

> Since $\dfrac{1}{5}$ is smaller than $\dfrac{3}{5}$, we cannot subtract until we borrow:
> $4\dfrac{1}{5} = 3 + \dfrac{5}{5} + \dfrac{1}{5} = 3 + \dfrac{6}{5} = 3\dfrac{6}{5}$

19. The LCD is 10.

$6 \boxed{\dfrac{3}{5} \cdot \dfrac{2}{2}} = \; 6 \dfrac{6}{10}$

$- \; 2 \boxed{\dfrac{1}{2} \cdot \dfrac{5}{5}} = - \; 2 \dfrac{5}{10}$

$4 \dfrac{1}{10}$

21. The LCD is 24.

$34 \boxed{\dfrac{1}{3} \cdot \dfrac{8}{8}} = \; 34 \dfrac{8}{24} = \; 33 \dfrac{32}{24}$

$- \; 12 \boxed{\dfrac{5}{8} \cdot \dfrac{3}{3}} = - \; 12 \dfrac{15}{24} = - \; 12 \dfrac{15}{24}$

$21 \dfrac{17}{24}$

Since $\dfrac{8}{24}$ is smaller than $\dfrac{15}{24}$, we cannot subtract until we borrow:

$34 \dfrac{8}{24} = 33 + \dfrac{24}{24} + \dfrac{8}{24} = 33 + \dfrac{32}{24} = 33 \dfrac{32}{24}$

23. $21 \qquad = \; 20 \dfrac{4}{4} \qquad \left(21 = 20 + 1 = 20 + \dfrac{4}{4} = 20\dfrac{4}{4} \right)$

$- \; 8 \dfrac{3}{4} = - \; 8 \dfrac{3}{4}$

$12 \dfrac{1}{4}$

25. $34 \qquad = \; 33 \dfrac{8}{8} \qquad \left(34 = 33 + 1 = 33 + \dfrac{8}{8} = 33\dfrac{8}{8} \right)$

$- \; 18 \dfrac{5}{8} = - \; 18 \dfrac{5}{8}$

$15 \dfrac{3}{8}$

27. $21 \boxed{\dfrac{1}{6} \cdot \dfrac{2}{2}} = \; 21 \dfrac{2}{12} = \; 20 \dfrac{14}{12}$

$- \; 13 \boxed{\dfrac{3}{4} \cdot \dfrac{3}{3}} = - \; 13 \dfrac{9}{12} = - \; 13 \dfrac{9}{12}$

$7 \dfrac{5}{12}$

Since $\dfrac{2}{12}$ is smaller than $\dfrac{9}{12}$, we cannot subtract until we borrow:

$21\dfrac{2}{12} = 20 + \dfrac{12}{12} + \dfrac{2}{12} = 20 + \dfrac{14}{12} = 20\dfrac{14}{12}$

29. The LCD is 8.

$14 \dfrac{1}{8} \qquad = \; 14 \dfrac{1}{8} = \; 13 \dfrac{9}{8}$

$- \boxed{\dfrac{3}{4} \cdot \dfrac{2}{2}} = - \; \dfrac{6}{8} = - \; \dfrac{6}{8}$

$13 \dfrac{3}{8}$

Since $\dfrac{1}{8}$ is smaller than $\dfrac{6}{8}$, we cannot subtract until we borrow:

$14\dfrac{1}{8} = 13 + \dfrac{8}{8} + \dfrac{1}{8} = 13 + \dfrac{9}{8} = 13\dfrac{9}{8}$

31. The LCD is 18.

$$25 \boxed{\frac{1}{9} \cdot \frac{2}{2}} = 25 \frac{2}{18} = 24 \frac{20}{18}$$

$$- 13 \boxed{\frac{5}{6} \cdot \frac{3}{3}} = -13 \frac{15}{18} = -13 \frac{15}{18}$$

$$11 \frac{5}{18}$$

Since $\frac{2}{18}$ is smaller than $\frac{15}{18}$, we cannot subtract until we borrow:

$$25 \frac{2}{18} = 24 + \frac{18}{18} + \frac{2}{18} = 24 \frac{20}{18}$$

33. Familiarize. We let w = the total weight of the meat.

Translate. We write an equation.

Weight of one package	plus	Weight of second package	is	Total weight
↓	↓	↓	↓	↓
$1\frac{1}{3}$	+	$4\frac{3}{5}$	=	w

Solve. To solve the equation we carry out the addition. The LCD is 15.

$$1 \boxed{\frac{1}{3} \cdot \frac{5}{5}} = 1 \frac{5}{15}$$

$$+ 4 \boxed{\frac{3}{5} \cdot \frac{3}{3}} = + 4 \frac{9}{15}$$

$$5 \frac{14}{15}$$

Check. We repeat the calculation. We also note that the answer is larger than either of the individual weights, so the answer seems reasonable.

State. The total weight of the meat was $5\frac{14}{15}$ lb.

35. Familiarize. Let h = the woman's excess height.

Translate. We have a "how much more" situation.

Height of son	plus	How much more height	is	Height of woman
↓	↓	↓	↓	↓
$150\frac{7}{10}$	+	h	=	$168\frac{1}{4}$

Solve. We solve the equation as follows:

$$h = 168 \frac{1}{4} - 150 \frac{7}{10}$$

$$168 \boxed{\frac{1}{4} \cdot \frac{5}{5}} = 168 \frac{5}{20} = 167 \frac{25}{20}$$

$$- 150 \boxed{\frac{7}{10} \cdot \frac{2}{2}} = -150 \frac{14}{20} = -150 \frac{14}{20}$$

$$17 \frac{11}{20}$$

35. (continued)

Check. We add the woman's excess height to her son's height:

$$17\frac{11}{20} + 150\frac{7}{10} = 17\frac{11}{20} + 150\frac{14}{20} = 167\frac{25}{20} =$$

$$168\frac{5}{20} = 168\frac{1}{4}$$

This checks.

State. The woman is $17\frac{11}{20}$ cm taller.

37. Familiarize. We let y = the length of the pencil.

Translate. We write an addition sentence.

Length of wood	+	Length of eraser	=	Total length
↓		↓		↓
$16\frac{9}{10}$	+	$1\frac{9}{10}$	=	y

Solve. To solve the equation we carry out the addition.

$$16 \frac{9}{10}$$

$$+ 1 \frac{9}{10}$$

$$17 \frac{18}{10} = 17 + \frac{18}{10} = 17 + 1\frac{8}{10} = 18\frac{8}{10} = 18\frac{4}{5}$$

Check. We repeat the calculation. We also note that the total length is larger than either of the individual lengths, so the answer seems reasonable.

State. The length of the standard pencil is $18\frac{4}{5}$ cm.

39. Familiarize. We draw a picture. We let D = the distance from Los Angeles.

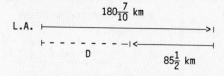

Translate. We write an equation.

Distance away from Los Angeles	minus	Distance toward Los Angeles	is	Distance from Los Angeles
↓	↓	↓	↓	↓
$180\frac{7}{10}$	-	$85\frac{1}{2}$	=	D

Solve. To solve the equation we carry out the subtraction. The LCM is 10.

$$180 \frac{7}{10} = 180 \frac{7}{10}$$

$$- 85 \boxed{\frac{1}{2} \cdot \frac{5}{5}} = -85 \frac{5}{10}$$

$$95 \frac{2}{10} = 95\frac{1}{5}$$

39. (continued)

Check. We add the distance from Los Angles to the distance the person drove toward Los Angeles:

$$95\frac{1}{5} + 85\frac{1}{2} = 95\frac{2}{10} + 85\frac{5}{10} = 180\frac{7}{10}$$

This checks.

State. The person was $95\frac{1}{5}$ km from Los Angeles.

41. Familiarize. We draw a picture.

$8\frac{1}{2}$ in.

11 in.   11 in.

$8\frac{1}{2}$ in.

Translate. We let D = the distance around the paper.

Top distance	plus	Right-side distance	plus	Bottom distance	plus	Left-side distance	is	Total distance
$8\frac{1}{2}$	+	11	+	$8\frac{1}{2}$	+	11	=	D

Solve. To solve we carry out the addition.

$$\begin{array}{r} 8\frac{1}{2} \\ 11 \\ 8\frac{1}{2} \\ + 11 \\ \hline 38\frac{2}{2} = 38 + 1 = 39 \end{array}$$

Check. We repeat the calculation.

State. The distance around the paper is 39 in.

43. Familiarize. We let c = the closing price.

Translate. We write an equation.

Amount at opening	-	Amount dropped	=	Amount at closing
$104\frac{5}{8}$	-	$1\frac{1}{4}$	=	c

Solve. To solve we carry out the subtraction. The LCD is 8.

$$\begin{array}{rcl} 104\frac{5}{8} & = & 104\frac{5}{8} \\ - 1\boxed{\frac{1}{4} \cdot \frac{2}{2}} & = & - 1\frac{2}{8} \\ \hline & & 103\frac{3}{8} \end{array}$$

43. (continued)

Check. We add the amount the stock dropped to the closing price:

$$1\frac{1}{4} + 103\frac{3}{8} = 1\frac{2}{8} + 103\frac{3}{8} = 104\frac{5}{8}$$

This checks.

State. The closing price was $103\frac{3}{8}$.

45. Familiarize. We let t = the total amount of paint used.

Translate. We write an equation.

Family room paint	+	Bedroom paint	=	Total paint
$1\frac{2}{3}$	+	$1\frac{1}{2}$	=	t

Solve. To solve we carry out the addition. The LCD is 6.

$$\begin{array}{rcl} 1\boxed{\frac{2}{3} \cdot \frac{2}{2}} & = & 1\frac{4}{6} \\ + 1\boxed{\frac{1}{2} \cdot \frac{3}{3}} & = & + 1\frac{3}{6} \\ \hline 2\frac{7}{6} & = & 3\frac{1}{6} \end{array}$$

Check. We repeat the calculation. We also note that the total is larger than either of the individual amounts, so the answer seems reasonable.

State. In all, $3\frac{1}{6}$ gal of paint were used.

47. Familiarize. We let h = the man's height.

Translate. We write an equation.

Son's height	+	Additional height of man	=	Man's height
$182\frac{9}{10}$	+	$5\frac{1}{4}$	=	h

Solve. To solve we carry out the addition. The LCD is 20.

$$\begin{array}{rcl} 182\boxed{\frac{9}{10} \cdot \frac{2}{2}} & = & 182\frac{18}{20} \\ + 5\boxed{\frac{1}{4} \cdot \frac{5}{5}} & = & + 5\frac{5}{20} \\ \hline 187\frac{23}{20} & = & 188\frac{3}{20} \end{array}$$

Check. We repeat the calculation. We also note that the man's height is larger than his son's height, so the answer seems reasonable.

State. The man is $188\frac{3}{20}$ cm tall.

49. <u>Familiarize</u>. We make a drawing. We let t = the total amount of paint needed and p = the additional paint needed.

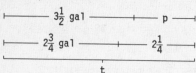

<u>Translate</u>. From the drawing we see that the additional paint needed is the total amount needed minus the amount the painter already has. Thus,

$$p = t - 3\tfrac{1}{2} = \left(2\tfrac{3}{4} + 2\tfrac{1}{4}\right) - 3\tfrac{1}{2}.$$

<u>Solve</u>. This is a two-step problem.

a) We first add $2\tfrac{3}{4}$ to $2\tfrac{1}{4}$ to find the total amount t of paint needed.

$$2\tfrac{3}{4}$$
$$+ \, 2\tfrac{1}{4}$$
$$\overline{}$$
$$4\tfrac{4}{4} = 5 = t$$

b) Next we subtract $3\tfrac{1}{2}$ from 5 to find the amount p of additional paint needed.

$$5 \;=\; 4\tfrac{2}{2}$$
$$-\,3\tfrac{1}{2} = -\,3\tfrac{1}{2}$$
$$\overline{}$$
$$1\tfrac{1}{2}$$

<u>Check</u>. We repeat the calculations.

<u>State</u>. The painter needed $1\tfrac{1}{2}$ gal of paint.

51. The length of each of the five sides is $5\tfrac{3}{4}$ yd. We add to find the distance around the figure.

$$5\tfrac{3}{4} + 5\tfrac{3}{4} + 5\tfrac{3}{4} + 5\tfrac{3}{4} + 5\tfrac{3}{4} = 25\tfrac{15}{4} = 25 + 3\tfrac{3}{4} = 28\tfrac{3}{4}$$

The distance is $28\tfrac{3}{4}$ yd.

53. We see that d and the two smallest distances combined are the same as the largest distance. We translate and solve.

$$2\tfrac{3}{4} + d + 2\tfrac{3}{4} = 12\tfrac{7}{8}$$
$$d = 12\tfrac{7}{8} - 2\tfrac{3}{4} - 2\tfrac{3}{4}$$
$$= 10\tfrac{1}{8} - 2\tfrac{3}{4} \qquad \text{Subtracting } 2\tfrac{3}{4}$$
$$\text{from } 12\tfrac{7}{8}$$
$$= 7\tfrac{3}{8} \qquad \text{Subtracting } 2\tfrac{3}{4} \text{ from}$$
$$10\tfrac{1}{8}$$

The length of d is $7\tfrac{3}{8}$ ft.

55. <u>Familiarize</u>. We let b = the length of the bolt.

<u>Translate</u>. From the drawing we see that the length of the small bolt is the sum of the diameters of the two tubes and the thicknesses of the two washers and the nut. Thus, we have $b = \tfrac{1}{2} + \tfrac{1}{16} + \tfrac{3}{4} + \tfrac{1}{16} + \tfrac{3}{16}.$

<u>Solve</u>. We carry out the addition. The LCD is 16.

$$b = \tfrac{1}{2} + \tfrac{1}{16} + \tfrac{3}{4} + \tfrac{1}{16} + \tfrac{3}{16} =$$
$$\tfrac{1}{2} \cdot \tfrac{8}{8} + \tfrac{1}{16} + \tfrac{3}{4} \cdot \tfrac{4}{4} + \tfrac{1}{16} + \tfrac{3}{16} =$$
$$\tfrac{8}{16} + \tfrac{1}{16} + \tfrac{12}{16} + \tfrac{1}{16} + \tfrac{3}{16} = \tfrac{25}{16} = 1\tfrac{9}{16}$$

<u>Check</u>. We repeat the calculation.

<u>State</u>. The smallest bolt is $1\tfrac{9}{16}$ in. long.

57. $\dfrac{12}{25} \div \dfrac{24}{5} = \dfrac{12}{25} \cdot \dfrac{5}{24} = \dfrac{12 \cdot 5}{25 \cdot 24} = \dfrac{12 \cdot 5 \cdot 1}{5 \cdot 5 \cdot 12 \cdot 2} =$
$\dfrac{12 \cdot 5}{12 \cdot 5} \cdot \dfrac{1}{5 \cdot 2} = \dfrac{1}{10}$

59. $$47\tfrac{2}{3} + n = 56\tfrac{1}{4}$$
$$47\tfrac{2}{3} + n - 47\tfrac{2}{3} = 56\tfrac{1}{4} - 47\tfrac{2}{3} \qquad \text{Subtracting } 47\tfrac{2}{3}$$
$$n + 0 = 56\tfrac{3}{12} - 47\tfrac{8}{12} \qquad \text{The LCD is 12.}$$
$$n = 55\tfrac{15}{12} - 47\tfrac{8}{12} \qquad \text{Borrowing}$$
$$n = 8\tfrac{7}{12}$$

Exercise Set 3.6

1. $8 \cdot 2\tfrac{5}{6}$

$= \dfrac{8}{1} \cdot \dfrac{17}{6}$ Writing fractional notation

$= \dfrac{8 \cdot 17}{1 \cdot 6} = \dfrac{2 \cdot 4 \cdot 17}{1 \cdot 2 \cdot 3} = \dfrac{2}{2} \cdot \dfrac{4 \cdot 17}{1 \cdot 3} = \dfrac{68}{3} = 22\tfrac{2}{3}$

3. $3\tfrac{5}{8} \cdot \tfrac{2}{3}$

$= \dfrac{29}{8} \cdot \dfrac{2}{3}$ Writing fractional notation

$= \dfrac{29 \cdot 2}{8 \cdot 3} = \dfrac{29 \cdot 2}{2 \cdot 4 \cdot 3} = \dfrac{2}{2} \cdot \dfrac{29}{4 \cdot 3} = \dfrac{29}{12} = 2\tfrac{5}{12}$

5. $3\tfrac{1}{2} \cdot 2\tfrac{1}{3} = \dfrac{7}{2} \cdot \dfrac{7}{3} = \dfrac{49}{6} = 8\tfrac{1}{6}$

7. $3\tfrac{2}{5} \cdot 2\tfrac{7}{8} = \dfrac{17}{5} \cdot \dfrac{23}{8} = \dfrac{391}{40} = 9\tfrac{31}{40}$

9. $4\tfrac{7}{10} \cdot 5\tfrac{3}{10} = \dfrac{47}{10} \cdot \dfrac{53}{10} = \dfrac{2491}{100} = 24\tfrac{91}{100}$

11. $20\tfrac{1}{2} \cdot 10\tfrac{1}{5} = \dfrac{41}{2} \cdot \dfrac{51}{5} = \dfrac{2091}{10} = 209\tfrac{1}{10}$

13. $20 \div 3\frac{1}{5}$

$= 20 \div \frac{16}{5}$ Writing fractional notation

$= 20 \cdot \frac{5}{16}$ Multiplying by the reciprocal

$= \frac{20 \cdot 5}{16} = \frac{4 \cdot 5 \cdot 5}{4 \cdot 4} = \frac{4}{4} \cdot \frac{5 \cdot 5}{4} = \frac{25}{4} = 6\frac{1}{4}$

15. $8\frac{2}{5} \div 7$

$= \frac{42}{5} \div 7$ Writing fractional notation

$= \frac{42}{5} \cdot \frac{1}{7}$ Multiplying by the reciprocal

$= \frac{42 \cdot 1}{5 \cdot 7} = \frac{6 \cdot 7}{5 \cdot 7} = \frac{7}{7} \cdot \frac{6}{5} = \frac{6}{5} = 1\frac{1}{5}$

17. $4\frac{3}{4} \div 1\frac{1}{3} = \frac{19}{4} \div \frac{4}{3} = \frac{19}{4} \cdot \frac{3}{4} = \frac{19 \cdot 3}{4 \cdot 4} = \frac{57}{16} = 3\frac{9}{16}$

19. $1\frac{7}{8} \div 1\frac{2}{3} = \frac{15}{8} \div \frac{5}{3} = \frac{15}{8} \cdot \frac{3}{5} = \frac{15 \cdot 3}{8 \cdot 5} = \frac{5 \cdot 3 \cdot 3}{8 \cdot 5}$

$= \frac{5}{5} \cdot \frac{3 \cdot 3}{8} = \frac{3 \cdot 3}{8} = \frac{9}{8} = 1\frac{1}{8}$

21. $5\frac{1}{10} \div 4\frac{3}{10} = \frac{51}{10} \div \frac{43}{10} = \frac{51}{10} \cdot \frac{10}{43} = \frac{51 \cdot 10}{10 \cdot 43}$

$= \frac{10}{10} \cdot \frac{51}{43} = \frac{51}{43} = 1\frac{8}{43}$

23. $20\frac{1}{4} \div 90 = \frac{81}{4} \div 90 = \frac{81}{4} \cdot \frac{1}{90} = \frac{81 \cdot 1}{4 \cdot 90} = \frac{9 \cdot 9 \cdot 1}{4 \cdot 9 \cdot 10}$

$= \frac{9}{9} \cdot \frac{9 \cdot 1}{4 \cdot 10} = \frac{9}{40}$

25. Familiarize. Visualize this situation as a rectangular array with $33\frac{1}{3}$ revolutions in each row and 21 rows.

Translate. We write an equation.

Revolutions per minute	$\cdot$	Number of minutes played	$=$	Total number of revolutions
$\downarrow$	$\downarrow$	$\downarrow$	$\downarrow$	$\downarrow$
$33\frac{1}{3}$	$\cdot$	21	$=$	n

Solve. To solve the equation we carry out the multiplication.

$n = 33\frac{1}{3} \cdot 21 = \frac{100}{3} \cdot 21 = \frac{100 \cdot 3 \cdot 7}{3 \cdot 1}$

$= \frac{3}{3} \cdot \frac{100 \cdot 7}{1} = 700$

Check. We repeat the calculation. We also note that $33\frac{1}{3} \approx 33$ and $21 \approx 20$. Then the product is about 660. Our answer seems reasonable.

State. The record makes 700 revolutions in 21 minutes.

27. Familiarize. We let n = the number of ounces of meat required.

Translate. We write an equation.

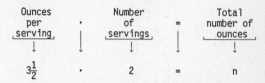

Ounces per serving	$\cdot$	Number of servings	$=$	Total number of ounces
$\downarrow$	$\downarrow$	$\downarrow$	$\downarrow$	$\downarrow$
$3\frac{1}{2}$	$\cdot$	2	$=$	n

Solve. To solve the equation we carry out the multiplication.

$n = 3\frac{1}{2} \cdot 2$

$= \frac{7}{2} \cdot \frac{2}{1} = \frac{7 \cdot 2}{2 \cdot 1} = \frac{2}{2} \cdot \frac{7}{1} = 7$

Check. We repeat the calculation.

State. 7 ounces of meat are needed.

29. Familiarize. We let w = the weight of $5\frac{1}{2}$ cubic feet of water.

Translate. We write an equation.

Weight per cubic foot	$\cdot$	Number of cubic feet	$=$	Total weight
$\downarrow$	$\downarrow$	$\downarrow$	$\downarrow$	$\downarrow$
$62\frac{1}{2}$	$\cdot$	$5\frac{1}{2}$	$=$	w

Solve. To solve the equation, we carry out the multiplication.

$w = 62\frac{1}{2} \cdot 5\frac{1}{2}$

$= \frac{125}{2} \cdot \frac{11}{2} = \frac{125 \cdot 11}{2 \cdot 2}$

$= \frac{1375}{4} = 343\frac{3}{4}$

Check. We repeat the calculation. We also note that $62\frac{1}{2} \approx 60$ and $5\frac{1}{2} \approx 5$. Then the product is about 300. Our answer seems reasonable.

State. The weight of $5\frac{1}{2}$ cubic feet of water is $343\frac{3}{4}$ lb.

31. Familiarize, Translate, and Solve. To find the ingredients for $\frac{1}{2}$ the recipe, we multiply each ingredient by $\frac{1}{2}$.

$$2\frac{1}{2} \cdot \frac{1}{2} = \frac{5}{2} \cdot \frac{1}{2} = \frac{5 \cdot 1}{2 \cdot 2} = \frac{5}{4} = 1\frac{1}{4}$$

$$1\frac{1}{3} \cdot \frac{1}{2} = \frac{4}{3} \cdot \frac{1}{2} = \frac{2 \cdot 2 \cdot 1}{3 \cdot 2} = \frac{2}{2} \cdot \frac{2 \cdot 1}{3} = \frac{2}{3}$$

$$\frac{2}{3} \cdot \frac{1}{2} = \frac{2 \cdot 1}{3 \cdot 2} = \frac{2}{2} \cdot \frac{1}{3} = \frac{1}{3}$$

$$\frac{1}{2} \cdot \frac{1}{2} = \frac{1 \cdot 1}{2 \cdot 2} = \frac{1}{4}$$

$$4 \cdot \frac{1}{2} = \frac{4 \cdot 1}{2} = \frac{4}{2} = 2$$

$$2 \cdot \frac{1}{2} = \frac{2 \cdot 1}{2} = \frac{2}{2} = 1$$

Check. We repeat the calculations.

State. The ingredients for $\frac{1}{2}$ the recipe are $1\frac{1}{4}$ lb opossum meat, $1\frac{1}{4}$ teaspoons salt, $\frac{2}{3}$ teaspoon black pepper, $\frac{1}{3}$ cup flour, $\frac{1}{4}$ cup water, 2 medium sweet potatoes, and 1 tablespoon sugar.

Familiarize, Translate, and Solve. To find the ingredients for 3 recipes, we multiply each ingredient by 3.

$$2\frac{1}{2} \cdot 3 = \frac{5}{2} \cdot 3 = \frac{5 \cdot 3}{2} = \frac{15}{2} = 7\frac{1}{2}$$

$$1\frac{1}{3} \cdot 3 = \frac{4}{3} \cdot 3 = \frac{4 \cdot 3}{3} = \frac{12}{3} = 4$$

$$\frac{2}{3} \cdot 3 = \frac{2 \cdot 3}{3} = \frac{6}{3} = 2$$

$$\frac{1}{2} \cdot 3 = \frac{1 \cdot 3}{2} = \frac{3}{2} = 1\frac{1}{2}$$

$$4 \cdot 3 = 12$$

$$2 \cdot 3 = 6$$

Check. We repeat the calculations.

State. The ingredients for 3 recipes are $7\frac{1}{2}$ lb opossum meat, $7\frac{1}{2}$ teaspoons salt, 4 teaspoons black pepper, 2 cups flour, $1\frac{1}{2}$ cups water, 12 medium sweet potatoes, and 6 tablespoons sugar.

33. Familiarize. We let t = the Fahrenheit temperature.
Translate.

Celsius temperature	times	$1\frac{4}{5}$	plus	32°	is	Fahrenheit temperature
↓	↓	↓	↓	↓	↓	↓
20	·	$1\frac{4}{5}$	+	32	=	t

Solve. We multiply and then add, according to the rules for order of operations.

$$t = 20 \cdot 1\frac{4}{5} + 32 = \frac{20}{1} \cdot \frac{9}{5} + 32 = \frac{20 \cdot 9}{1 \cdot 5} + 32 =$$

$$\frac{4 \cdot 5 \cdot 9}{1 \cdot 5} + 32 = \frac{5}{5} \cdot \frac{4 \cdot 9}{1} + 32 = 36 + 32 = 68$$

Check. We repeat the calculation.
State. 68° Fahrenheit corresponds to 20° Celsius.

35. Familiarize. Visualize a rectangular array containing 480 revolutions with $33\frac{1}{3}$ revolutions in each row. We must determine how many rows the array has.
Translate. The division that corresponds to the situation is

$$480 \div 33\frac{1}{3} = t.$$

Solve. To solve the equation we carry out the division.

$$t = 480 \div 33\frac{1}{3} = 480 \div \frac{100}{3}$$

$$= 480 \cdot \frac{3}{100} = \frac{20 \cdot 24 \cdot 3}{20 \cdot 5}$$

$$= \frac{20}{20} \cdot \frac{24 \cdot 3}{5} = \frac{72}{5}$$

$$= 14\frac{2}{5}$$

Check. We repeat the calculation. We also note that $33\frac{1}{3} \approx 30$. Then the quotient is about 16. Our answer seems reasonable.

State. The record played $14\frac{2}{5}$ minutes.

37. Familiarize. We let m = the number of miles per gallon the car got.
Translate. We write an equation.

Total number of miles traveled	÷	Number of gallons of gas used	=	Miles per gallon
↓	↓	↓	↓	↓
213	÷	$14\frac{2}{10}$	=	m

Solve. To solve the equation we carry out the division.

$$m = 213 \div 14\frac{2}{10} = 213 \div \frac{142}{10}$$

$$= 213 \cdot \frac{10}{142} = \frac{3 \cdot 71 \cdot 2 \cdot 5}{2 \cdot 71 \cdot 1}$$

$$= \frac{2 \cdot 71}{2 \cdot 71} \cdot \frac{3 \cdot 5}{1} = 15$$

Check. We repeat the calculation.
State. Thus, the car got 15 miles per gallon of gas.

39. **Familiarize.** We let n = the number of cubic feet occupied by 250 lb of water.

Translate. We write an equation.

Total weight	÷	Weight per cubic foot	=	Number of cubic feet
↓	↓	↓		↓
250	÷	$62\frac{1}{2}$	=	n

Solve. To solve the equation we carry out the division.

$$n = 250 \div 62\frac{1}{2} = 250 \div \frac{125}{2}$$

$$= 250 \cdot \frac{2}{125} = \frac{2 \cdot 125 \cdot 2}{125 \cdot 1}$$

$$= \frac{125}{125} \cdot \frac{2 \cdot 2}{1} = 4$$

Check. We repeat the calculation.

State. 4 cubic feet would be occupied.

41. **Familiarize.** We let p = the number of pounds of turkey needed for 32 servings.

Translate. We write an equation.

Total number of servings	÷	Servings per pound	=	Number of pounds needed
↓	↓	↓		↓
32	÷	$1\frac{1}{3}$	=	p

Solve. To solve the equation we carry out the division.

$$p = 32 \div 1\frac{1}{3} = 32 \div \frac{4}{3}$$

$$= 32 \cdot \frac{3}{4} = \frac{4 \cdot 8 \cdot 3}{4 \cdot 1}$$

$$= \frac{4}{4} \cdot \frac{8 \cdot 3}{1} = 24$$

Check. We repeat the calculation.

State. 24 pounds would be needed.

43. **Familiarize.** The figure is composed of two rectangles. One has dimensions s by $\frac{1}{2} \cdot$ s, or $6\frac{7}{8}$ in. by $\frac{1}{2} \cdot 6\frac{7}{8}$ in. The other has dimensions $\frac{1}{2} \cdot$ s by $\frac{1}{2} \cdot$ s, or $\frac{1}{2} \cdot 6\frac{7}{8}$ in. by $\frac{1}{2} \cdot 6\frac{7}{8}$ in. The total area is the sum of the areas of these two rectangles. We let A = the total area.

Translate. We write an equation.

$$A = \left(6\frac{7}{8}\right) \cdot \left(\frac{1}{2} \cdot 6\frac{7}{8}\right) + \left(\frac{1}{2} \cdot 6\frac{7}{8}\right) \cdot \left(\frac{1}{2} \cdot 6\frac{7}{8}\right)$$

43. (continued)

Solve. We carry out each multiplication and then add.

$$A = \left(6\frac{7}{8}\right) \cdot \left(\frac{1}{2} \cdot 6\frac{7}{8}\right) + \left(\frac{1}{2} \cdot 6\frac{7}{8}\right) \cdot \left(\frac{1}{2} \cdot 6\frac{7}{8}\right)$$

$$= \frac{55}{8} \cdot \left(\frac{1}{2} \cdot \frac{55}{8}\right) + \left(\frac{1}{2} \cdot \frac{55}{8}\right) \cdot \left(\frac{1}{2} \cdot \frac{55}{8}\right)$$

$$= \frac{55}{8} \cdot \frac{55}{16} + \frac{55}{16} \cdot \frac{55}{16}$$

$$= \frac{3025}{128} + \frac{3025}{256} = \frac{3025}{128} \cdot \frac{2}{2} + \frac{3025}{256}$$

$$= \frac{6050}{256} + \frac{3025}{256} = \frac{9075}{256}$$

$$= 35\frac{115}{256}$$

Check. We repeat the calculation.

State. The area is $35\frac{115}{256}$ sq in.

45. **Familiarize.** We make a drawing.

$302\frac{1}{2}$ m

Translate. We let A = the area of the lot not covered by the building.

Area left over	is	Area of lot	minus	Area of building
↓	↓	↓	↓	↓
A	=	$\left(302\frac{1}{2}\right) \cdot \left(205\frac{1}{4}\right)$	−	$(100) \cdot \left(25\frac{1}{2}\right)$

Solve. We do each multiplication and then find the difference.

$$A = \left(302\frac{1}{2}\right) \cdot \left(205\frac{1}{4}\right) - (100) \cdot \left(25\frac{1}{2}\right)$$

$$= \frac{605}{2} \cdot \frac{821}{4} - \frac{100}{1} \cdot \frac{51}{2}$$

$$= \frac{605 \cdot 821}{2 \cdot 4} - \frac{100 \cdot 51}{1 \cdot 2}$$

$$= \frac{605 \cdot 821}{2 \cdot 4} - \frac{2 \cdot 50 \cdot 51}{1 \cdot 2} = \frac{605 \cdot 821}{2 \cdot 4} - \frac{2}{2} \cdot \frac{50 \cdot 51}{1}$$

$$= \frac{496,705}{8} - 2550 = 62,088\frac{1}{8} - 2550$$

$$= 59,538\frac{1}{8}$$

Check. We repeat the calculation.

State. The area left over is $59,538\frac{1}{8}$ sq m.

47.
```
      6 7 0 9
    ×   2 1 3
    2 0 1 2 7
    6 7 0 9 0
  1 3 4 1 8 0 0
  1,4 2 9,0 1 7
```

49. $\frac{5}{7} \cdot t = 420$

$t = 420 \div \frac{5}{7} = 420 \cdot \frac{7}{5} = \frac{5 \cdot 84 \cdot 7}{5 \cdot 1} = \frac{5}{5} \cdot \frac{84 \cdot 7}{1}$

$= 588$

51. $8 \div \frac{1}{2} + \frac{3}{4} + \left(5 - \frac{5}{8}\right)^2 = 8 \div \frac{1}{2} + \frac{3}{4} + \left(\frac{40}{8} - \frac{5}{8}\right)^2 =$

$8 \div \frac{1}{2} + \frac{3}{4} + \left(\frac{35}{8}\right)^2 = 8 \div \frac{1}{2} + \frac{3}{4} + \frac{1225}{64} =$

$8 \cdot 2 + \frac{3}{4} + \frac{1225}{64} = 16 + \frac{3}{4} + \frac{1225}{64} =$

$\frac{1024}{64} + \frac{48}{64} + \frac{1225}{64} = \frac{2297}{64} = 35\frac{57}{64}$

53. $\frac{1}{3} \div \left(\frac{1}{2} - \frac{1}{5}\right) \times \frac{1}{4} + \frac{1}{6}$

$= \frac{1}{3} \div \left(\frac{5}{10} - \frac{2}{10}\right) \times \frac{1}{4} + \frac{1}{6}$

$= \frac{1}{3} \div \frac{3}{10} \times \frac{1}{4} + \frac{1}{6}$

$= \frac{1}{3} \times \frac{10}{3} \times \frac{1}{4} + \frac{1}{6}$

$= \frac{10}{9} \times \frac{1}{4} + \frac{1}{6}$

$= \frac{2 \times 5 \times 1}{9 \times 2 \times 2} + \frac{1}{6} = \frac{2}{2} \times \frac{5 \times 1}{9 \times 2} + \frac{1}{6}$

$= \frac{5}{18} + \frac{1}{6} = \frac{5}{18} + \frac{3}{18} = \frac{8}{18} = \frac{4}{9}$

55. $4\frac{1}{2} \div 2\frac{1}{2} + 8 - 4 \div \frac{1}{2}$

$= \frac{9}{2} \div \frac{5}{2} + 8 - 4 \div \frac{1}{2}$

$= \frac{9}{2} \cdot \frac{2}{5} + 8 - 4 \div \frac{1}{2}$

$= \frac{2}{2} \cdot \frac{9}{5} + 8 - 4 \div \frac{1}{2}$

$= \frac{9}{5} + 8 - 4 \div \frac{1}{2}$

$= \frac{9}{5} + 8 - 4 \cdot 2$

$= \frac{9}{5} + 8 - 8$

$= \frac{9}{5} + \frac{40}{5} - \frac{40}{5} = \frac{9}{5}$, or $1\frac{4}{5}$

Exercise Set 4.1

1. a) Write a word name
   for the whole number.   ┌──────────────┐
                            │ Twenty-three │
                            └──────────────┘

   b) Write "and" for the   Twenty-three
      decimal point.   ┌─────┐
                               │ and │
                               └─────┘

   c) Write a word name
      for the number to   Twenty-three
      the right of the   and
      decimal point,
      followed by the   ┌───────────┐
      place value of the   │ two tenths │
      last digit.   └───────────┘

   A word name for 23.2 is twenty-three and two
   tenths.

3.   1 3 5 . 8 7

   One hundred thirty-five and eighty-seven
   hundredths

5.   3 4 . 8 9 1

   Thirty-four and eight hundred ninety-one
   thousandths

7. Write "and 48 cents" as "and $\frac{48}{100}$ dollars." A word
   name for $326.48 is three hundred twenty-six and
   $\frac{48}{100}$ dollars.

9. Write "and 67 cents" as "and $\frac{67}{100}$ dollars." A word
   name for $0.67 is zero and $\frac{67}{100}$ dollars.

11. 6.8   6.8. $\frac{68}{10}$

   1 place   Move 1 place   1 zero

        $6.8 = \frac{68}{10}$

13. 0.17   0.17. $\frac{17}{100}$

   2 places   Move 2 places   2 zeros

        $0.17 = \frac{17}{100}$

15. 1.46   1.46. $\frac{146}{100}$

   2 places   Move 2 places   2 zeros

        $1.46 = \frac{146}{100}$

17. 204.6   204.6. $\frac{2046}{10}$

   1 place   Move 1 place   1 zero

        $204.6 = \frac{2046}{10}$

19. 3.142   3.142. $\frac{3142}{1000}$

   3 places   Move 3 places   3 zeros

        $3.142 = \frac{3142}{1000}$

21. 46.03   46.03. $\frac{4603}{100}$

   2 places   Move 2 places   2 zeros

        $46.03 = \frac{4603}{100}$

23. 0.00013   0.00013. $\frac{13}{100,000}$

   5 places   Move 5 places   5 zeros

        $0.00013 = \frac{13}{100,000}$

25. 20.003   20.003. $\frac{20,003}{1000}$

   3 places   Move 3 places   3 zeros

        $20.003 = \frac{20,003}{1000}$

27. 1.0008   1.0008. $\frac{10008}{10000}$

   4 places   Move 4 places   4 zeros

        $1.0008 = \frac{10,008}{10,000}$

29. 4567.2   4567.2. $\frac{45672}{10}$

   1 place   Move 1 place   1 zero

        $4567.2 = \frac{45,672}{10}$

31. $\frac{8}{10}$ 0.8.

   1 zero   Move 1 place

        $\frac{8}{10} = 0.8$

33. $\frac{92}{100}$ 0.92.

   2 zeros   Move 2 places

        $\frac{92}{100} = 0.92$

35. $\dfrac{93}{10}$ 9.3.⌐

1 zero   Move 1 place

$\dfrac{93}{10} = 9.3$

37. $\dfrac{889}{100}$ 8.89.⌐

2 places   Move 2 places

$\dfrac{889}{100} = 8.89$

39. $\dfrac{2508}{10}$ 250.8.⌐

1 place   Move 1 place

$\dfrac{2508}{10} = 250.8$

41. $\dfrac{3798}{1000}$ 3.798.⌐

3 places   Move 3 places

$\dfrac{3798}{1000} = 3.798$

43. $\dfrac{78}{10,000}$ 0.0078.⌐

4 zeros   Move 4 places

$\dfrac{78}{10,000} = 0.0078$

45. $\dfrac{56,788}{100,000}$ 0.56788.⌐

5 zeros   Move 5 places

$\dfrac{56,788}{100,000} = 0.56788$

47. $\dfrac{2173}{100}$ 21.73.⌐

2 zeros   Move 2 places

$\dfrac{2173}{100} = 21.73$

49. $\dfrac{66}{100}$ 0.66.⌐

2 zeros   Move 2 places

$\dfrac{66}{100} = 0.66$

51. $\dfrac{3417}{100}$ 34.17.⌐

2 zeros   Move 2 places

$\dfrac{3417}{100} = 34.17$

53. $\dfrac{376,193}{1,000,000}$ 0.376193.⌐

6 zeros   Move 6 places

$\dfrac{376,193}{1,000,000} = 0.376193$

55. Round 617 ☐2 to the nearest ten.

The digit 7 is in the tens place. Since the next digit to the right (2) is 4 or lower, we round down. The answer is 6170.

57. Round 6 ☐1 72 to the nearest thousand.

Since 1 is 4 or lower we round down. The answer is 6000.

59. $4\dfrac{909}{1000} = \dfrac{4909}{1000}$ 4.909.⌐

3 zeros   Move 3 places

$4\dfrac{909}{1000} = 4.909$

Exercise Set 4.2

1. To compare two numbers in decimal notation, start at the <u>left</u> and compare corresponding digits. When the two digits differ, the number with the larger digit is the larger of the two numbers.

    0.06

    0.58   These digits differ, and 5 is larger than 0.

    Thus, 0.58 is larger.

3. 0.10 ⟵ Extra zero

    0.111   Starting at the left, these digits are the first to differ, and 1 is larger than 0.

    Thus, 0.111 is larger.

5. 0.0009

    0.001   Starting at the left, these digits are the first to differ, and 1 is larger than 0.

    Thus, 0.001 is larger.

7. 234.07
    ↑
    235.07

Starting at the left, these digits are the first to differ and 5 is larger than 4.

Thus, 235.07 is larger.

9. 0.4545
    ↑
    0.05454

These digits differ, and 4 is larger than 0.

Thus, 0.4545 is larger.

11. $\frac{4}{100}$ = 0.04 so we compare 0.004 and 0.04.

     0.004
     ↑
     0.04

Starting at the left, these digits are the first to differ, and 4 is larger than 0.

Thus, 0.04 or $\frac{4}{100}$ is larger.

13. 0.54
     ↑
     0.78

These digits differ, and 7 is larger than 5.

Thus, 0.78 is larger.

15. 0.8437
     ↑
     0.84384

Starting at the left, these digits are the first to differ, and 8 is larger than 7.

Thus, 0.84384 is larger.

17. 0.19
     ↑
     1.9

Starting at the left, these digits are the first to differ, and 1 is larger than 0.

Thus, 1.9 is larger.

19. 0.1⌐1⌐
     ↓
     0.1

Hundredths digit is less than 5. Round down.

21. 0.1⌐6⌐
     ↓
     0.2

Hundredths digit is 5 or higher. Round up.

23. 0.5⌐7⌐94
     ↓
     0.6

Hundredths digit is 5 or higher. Round up.

25. 2.7⌐4⌐49
     ↓
     2.7

Hundredths digit is less than 5. Round down.

27. 13.4⌐1⌐
     ↓
     13.4

Hundredths digit is less than 5. Round down.

29. 123.6⌐5⌐
     ↓
     123.7

Hundredths digit is 5 or higher. Round up.

31. 0.89⌐3⌐
     ↓
     0.89

Thousandths digit is less than 5. Round down.

33. 0.66⌐6⌐
     ↓
     0.67

Thousandths digit is 5 or higher. Round up.

35. 0.42⌐4⌐6
     ↓
     0.42

Thousandths digit is less than 5. Round down.

37. 1.43⌐5⌐
     ↓
     1.44

Thousandths digit is 5 or higher. Round up.

39. 3.58⌐1⌐
     ↓
     3.58

Thousandths digit is less than 5. Round down.

41. 0.00⌐7⌐
     ↓
     0.01

Thousandths digit is 5 or higher. Round up.

43. 0.324⌐6⌐
     ↓
     0.325

Ten-thousandths digit is 5 or higher. Round up.

45. 0.666⌐6⌐
     ↓
     0.667

Ten-thousandths digit is 5 or higher. Round up.

47. 17.001⌐5⌐
     ↓
     17.002

Ten-thousandths digit is 5 or higher. Round up.

49. 0.000[9]   Ten-thousandths digit is 5 or higher.
    ↓   Round up.
    0.001

51. 10.101[1] Ten-thousandths digit is less than 5.
    ↓   Round down.
    10.101

53. 0.116[1]   Ten-thousandths digit is less than 5.
    ↓   Round down.
    0.116

55. 2[8]3.1359 Tens digit is 5 or higher. Round up.
    ↓
    300

57. 283.135[9] Ten-thousandths digit is 5 or higher.
    ↓   Round up.
    283.136

59. 283.[1]359 Tenths digit is less than 5. Round
    ↓   down.
    283

61. 34.5438[9] Hundred-thousandths digit is 5 or
    ↓   higher. Round up.
    34.5439

63. 34.54[3]89 Thousandths digit is less than 5.
    ↓   Round down.
    34.54

65. 34.[5]4389 Tenths digit is 5 or higher. Round
    ↓   up.
    35

67.   ¹ ¹
      6 8 1
    + 1 4 9
      8 3 0

69.   ¹ ¹⁶
        2 ⁄8 7
    -     8 5
        1 8 2

71. 6.78346[123] ← Drop all decimal places past the
    ↓   fifth place.
    6.78346

73. 99.99999[9999] ← Drop all decimal places past the
    ↓   fifth place.
    99.99999

Exercise Set 4.3

1. 3 ¹6.2 5   Add hundredths.
                 Add tenths.
    + 1 8.1 2   Write a decimal point in the
    3 3 4.3 7   answer.
                 Add ones.
                 Add tens.
                 Add hundreds.

3.   6 ⁵⁹.4 0 3   Add thousandths.
                   Add hundredths.
    +   9 1 6.8 1 2   Add tenths.
    1 5 7 6.2 1 5   Write a decimal point in the
                   answer.
                   Add ones.
                   Add tens.
                   Add hundreds.

5.   ¹ 9.1 0̇ 4
     + 1 2 3.4 5 6
     1 3 2.5 6 0

7.   6 1.0 0̇ 6
    +   3.4 0 7
      6 4.4 1 3

9.   2 0.0 1 2 4
    + 3 0.0 1 2 4
      5 0.0 2 4 8

11.   0.8 3 0 ← An extra 0 can be written here
    + 0.0 0 5   if desired.
      0.8 3 5   Adding

13. Line up the decimal places.

      ¹
      0.3 4 0   Writing an extra 0
      3.5 0 0   Writing 2 extra 0's
      0.1 2 7
    + 7 6 8.0 0 0   Writing in the decimal point
      7 7 1.9 6 7   and 3 extra 0's
                    Adding

15.   ¹ 7.0̇ 0̇ 0 0   Writing in the decimal point.
      3.2 4 0 0   You may find it helpful to
      0.2 5 6 0   write extra 0's.
    + 0.3 6 8 9
      2 0.8 6 4 9

17.   2.7 0 3 0
        7 8.3 3 0 0
        2 8.0 0 0 9
    + 1 1 8.4 3 4 1
      2 2 7.4 6 8 0

19.   ¹ ² ¹ ¹
        9 9.6 0 0 1
        7 2 8 5.1 8 0 0
        5 0 0.0 4 2 0
    +   8 7 0.0 0 0 0
        8 7 5 4.8 2 2 1

21.
$$\overset{4\ \ \overset{12}{}}{\cancel{5}.\cancel{2}}$$
$$-\ 3.9$$
$$\overline{1.3}$$

Borrow ones to subtract tenths.
Subtract tenths.
Write a decimal point in the answer.
Subtract ones.

23.
$$\overset{4\ \ 11\ \ 2\ \ 11}{\cancel{5}\cancel{1}.\cancel{3}\cancel{1}}$$
$$-\ 2.2\ 9$$
$$\overline{4\ 9.0\ 2}$$

Borrow tenths to subtract hundredths. Subtract hundredths.
Subtract tenths.
Write a decimal point in the answer.
Borrow tens to subtract ones. Subtract ones.
Subtract tens.

25.
$$4\ 8.7\ 6$$
$$-\ 3.1\ 5$$
$$\overline{4\ 5.6\ 1}$$

27.
$$\overset{8\ \cancel{2}\ \overset{11}{13}}{\cancel{9}\cancel{2}.\cancel{3}4\ 1}$$
$$-\ \ \ 6.4\ 2$$
$$\overline{8\ 5.9\ 2\ 1}$$

29.
$$\overset{4\ 9\ 9\ 10}{2.\cancel{5}\cancel{0}\cancel{0}\cancel{0}}$$
$$-\ 0.0\ 0\ 2\ 5$$
$$\overline{2.4\ 9\ 7\ 5}$$
Writing 3 extra 0's

31.
$$\overset{3\ 9\ 10}{3.\cancel{4}\cancel{0}\cancel{0}}$$
$$-\ 0.0\ 0\ 3$$
$$\overline{3.3\ 9\ 7}$$
Writing 2 extra 0's

33. Line up the decimal points. Write an extra 0 if desired.
$$\overset{1\ 7\ \overset{11}{}\ \ }{\cancel{2}\cancel{8}.\cancel{2}\cancel{0}}$$
$$-\ 1\ 9.3\ 5$$
$$\overline{8.8\ 5}$$

35.
$$3\ 4.0\ 7$$
$$-\ 3\ 0.7$$
$$\overline{3.3\ 7}$$

37.
$$\overset{\ \ \ \ \ 4\ 10}{8.4\ \cancel{5}\cancel{0}}$$
$$-\ 7.4\ 0\ 5$$
$$\overline{1.0\ 4\ 5}$$

39.
$$\overset{5\ 10}{\cancel{6}.\cancel{0}\ 0\ 3}$$
$$-\ 2.3$$
$$\overline{3.7\ 0\ 3}$$

41.
$$\overset{9\ 9\ 9\ 10}{\cancel{1}.\cancel{0}\cancel{0}\cancel{0}\cancel{0}}$$
$$-\ 0.0\ 0\ 9\ 8$$
$$\overline{0.9\ 9\ 0\ 2}$$
Writing in the decimal point and 4 extra 0's
Subtracting

43.
$$\overset{9\ 9\ 9\ 10}{\cancel{1}\cancel{0}\cancel{0}.\cancel{0}\cancel{0}}$$
$$-\ \ \ \ 0.3\ 4$$
$$\overline{9\ 9.6\ 6}$$

45.
$$\overset{6\ \overset{14}{}}{\cancel{7}.\cancel{4}\ 8}$$
$$-\ \ 2.6$$
$$\overline{4.8\ 8}$$

47.
$$\overset{\ \ \ \ 4\ 10}{2\ \cancel{5}.\cancel{0}\ 0\ 8}$$
$$-\ \ 1\ 2.4$$
$$\overline{1\ 2.6\ 0\ 8}$$

49.
$$\overset{\ \ \ \ \ \ \ \ \ \ 7\ 10}{2\ 5\ 4\ 8.9\ \cancel{8}\cancel{0}}$$
$$-\ \ \ \ \ \ \ \ 2.0\ 0\ 7$$
$$\overline{2\ 5\ 4\ 6.9\ 7\ 3}$$

51.
$$\overset{\ \ \ 4\ 9\ 9\ 10}{4\ \cancel{5}.\cancel{0}\cancel{0}\cancel{0}}$$
$$-\ \ 0.9\ 9\ 9$$
$$\overline{4\ 4.0\ 0\ 1}$$

53.
$$\overset{\ \ \ 8\ 10}{3.\cancel{9}\cancel{0}\ 7}$$
$$-\ 1.4\ 1\ 6$$
$$\overline{2.4\ 9\ 1}$$

55.
$$\overset{\ \ \ \ \ \ \ \ 8\ 17}{3\ 2.7\ \cancel{9}\cancel{7}\ 8}$$
$$-\ \ 0.0\ 5\ 9\ 2$$
$$\overline{3\ 2.7\ 3\ 8\ 6}$$

57.
$$\overset{2\ 9\ 10\ 6\ 14}{\cancel{3}.\cancel{0}\cancel{0}\cancel{7}\cancel{4}}$$
$$-\ 1.3\ 4\ 0\ 8$$
$$\overline{1.6\ 6\ 6\ 6}$$

59.
$$\overset{\ \ \ \ \ \ \ \ \ 4\ 8\ 9\ 17}{2\ 3\ 4\ \cancel{5}.\cancel{9}\cancel{0}\ 7\ 8\ 6}$$
$$-\ \ \ \ \ \ 0.9\ 9\ 9$$
$$\overline{2\ 3\ 4\ 4.9\ 0\ 8\ 8\ 6}$$

61. $x + 0.223 = 200.12$

$x = 200.12 - 0.223$ Subtracting 0.223 on both sides

$x = 199.897$
$$\overset{\ \ \ \ \ 1\ 9\ 9\ \overset{10\ 11}{\cancel{0}\ \cancel{1}}\ 10}{2\ 0\ 0.\cancel{1}\cancel{2}\cancel{0}}$$
$$-\ \ \ \ 0.2\ 2\ 3$$
$$\overline{1\ 9\ 9.8\ 9\ 7}$$

63. $3.205 + m = 22.456$

$m = 22.456 - 3.205$ Subtracting 3.205 on both sides

$m = 19.251$
$$\overset{1\ \overset{12}{\cancel{2}}}{2\ \cancel{2}.4\ 5\ 6}$$
$$-\ \ \ 3.2\ 0\ 5$$
$$\overline{1\ 9.2\ 5\ 1}$$

65. $17.95 + p = 402.63$

$p = 402.63 - 17.95$ Subtracting 17.95 on both sides

$p = 384.68$
$$\overset{3\ 9\ \overset{11\ 15}{\cancel{1}\ \cancel{5}}\ 13}{4\ \cancel{0}\ \cancel{2}.\cancel{6}\ 3}$$
$$-\ \ 1\ 7.9\ 5$$
$$\overline{3\ 8\ 4.6\ 8}$$

67. $13{,}083.3 = x + 12{,}500.33$

$13{,}083.3 - 12{,}500.33 = x$ Subtracting 12,500.33 on both sides

$582.97 = x$
$$\overset{\ \ \ \ \ \ 2\ 10\ \ \ 2\ \overset{12}{\cancel{2}}\ 10}{1\ \cancel{3}.\cancel{0}\ 8\ \cancel{3}.\cancel{3}\cancel{0}}$$
$$-\ 1\ 2{,}5\ 0\ 0.3\ 3$$
$$\overline{5\ 8\ 2.9\ 7}$$

69. 34,⎡5⎤67   Hundreds digit is 5 or higher.
       ↓   Round up.
    35,000

71. First, "undo" the incorrect addition by subtracting 235.7 from the incorrect answer:

$$\begin{array}{r} 8\ 1\ 7.2 \\ -\ 2\ 3\ 5.7 \\ \hline 5\ 8\ 1.5 \end{array}$$

The original minuend was 581.5. Now subtract 235.7 from this as the student originally intended:

$$\begin{array}{r} 5\ 8\ 1.5 \\ -\ 2\ 3\ 5.7 \\ \hline 3\ 4\ 5.8 \end{array}$$

The correct answer is 345.8.

Exercise Set 4.4

1. Familiarize. We visualize the situation. We let g = the number of gallons purchases.

23.6 gal	17.7 gal	20.8 gal	17.2 gal	25.4 gal	13.8 gal
g					

Translate. Amounts are being combined. We translate to an equation.

First, plus second, plus third, plus fourth,
↓   ↓   ↓   ↓
23.6 + 17.7 + 20.8 + 17.2

plus fifth, plus sixth, is total,
↓   ↓   ↓
+ 25.4 + 13.8 = g

Solve. To solve the equation we carry out the addition.

$$\begin{array}{r} \overset{2\ 3}{2}\ 3.6 \\ 1\ 7.7 \\ 2\ 0.8 \\ 1\ 7.2 \\ 2\ 5.4 \\ +\ 1\ 3.8 \\ \hline 1\ 1\ 8.5 \end{array}$$

Thus, g = 118.5.

Check. We can check by repeating the addition. We also note that the answer seems reasonable since it is larger than any of the numbers being added. We can also check by rounding:

23.6 + 17.7 + 20.8 + 17.2 + 25.4 + 13.8 ≈

24 + 18 + 21 + 17 + 25 + 14 = 119 ≈ 118.5

State. 118.5 gallons of gasoline were purchased.

3. Familiarize. We visualize the situation. We let c = the amount of change.

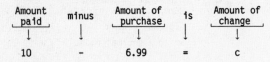

Translate. This is a "take-away" situation.

Amount paid	minus	Amount of purchase	is	Amount of change
↓	↓	↓	↓	↓
10	−	6.99	=	c

Solve. To solve the equation we carry out the subtraction.

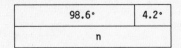

$$\begin{array}{r} \overset{9\ 9\ \ 10}{1\ \cancel{0}.\cancel{0}\ \cancel{0}} \\ -\ 6.9\ 9 \\ \hline 3.0\ 1 \end{array}$$

Thus, c = $3.01.

Check. We check by adding 3.01 to 6.99 to get 10. This checks.

State. The change was $3.01.

5. Familiarize. We visualize the situation. We let n = the new temperature.

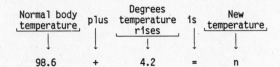

Translate. We are combining amounts.

Normal body temperature	plus	Degrees temperature rises	is	New temperature
↓	↓	↓	↓	↓
98.6	+	4.2	=	n

Solve. To solve we carry out the addition.

$$\begin{array}{r} \overset{1}{9}\ 8.6 \\ +\ \ \ \ 4.2 \\ \hline 1\ 0\ 2.8 \end{array}$$

Thus n = 102.8.

Check. We can check by repeating the addition. We can also check by rounding:

98.6 + 4.2 ≈ 99 + 4 = 103 ≈ 102.8.

State. The new temperature was 102.8° F.

7. <u>Familiarize</u>. We visualize the situation. We let m = the odometer reading at the end of the trip.

22,456.8 mi	234.7 mi
m	

<u>Translate</u>. We are combining amounts.

Reading before trip	plus	Miles driven	is	Reading at end of trip
↓	↓	↓	↓	↓
22,456.8	+	234.7	=	m

<u>Solve</u>. To solve we carry out the addition.

$$
\begin{array}{r}
2\,2,4\,\overset{1}{5}\,\overset{1}{6}.8 \\
+\quad 2\,3\,4.7 \\
\hline
2\,2,6\,9\,1.5
\end{array}
$$

Thus m = 22,691.5.

<u>Check</u>. We can check by repeating the addition. We can also check by rounding:

22,456.8 + 234.7 ≈ 22,460 + 230 = 22,690 ≈ 22,691.5

<u>State</u>. The odometer reading at the end of the trip was 22,691.5.

9. <u>Familiarize</u>. We visualize the situation. We let a = the amount by which the cost of a lunch in New York exceeds the cost in Los Angeles.

$20.27	
$13.68	a

<u>Translate</u>. This is a "how-much-more" situation.

Cost of lunch in L.A.	plus	Additional cost in N.Y.	is	Cost of lunch in N.Y.
↓	↓	↓	↓	↓
13.68	+	a	=	20.27

<u>Solve</u>. To solve the equation we subtract 13.68 on both sides.

a = 20.27 - 13.68

a = 6.59

$$
\begin{array}{r}
\overset{1}{\cancel{2}}\overset{9}{\cancel{0}}.\overset{1}{\cancel{2}}\overset{17}{\cancel{7}} \\
-\ 1\,3.6\,8 \\
\hline
6.5\,9
\end{array}
$$

<u>Check</u>. We can check by adding 6.59 to 13.68 to get 20.27. This checks.

<u>State</u>. Lunch is $6.59 higher in New York than in Los Angeles.

11. <u>Familiarize</u>. We visualize the situation. We let s = the total amount spent.

$26.79	$268.10	$690.00
	s	

<u>Translate</u>. We are combining amounts.

First check	+	Second check	+	Third check	is	Total spent
↓		↓		↓	↓	↓
26.79	+	268.10	+	690.00	=	s

<u>Solve</u>. To solve we carry out the addition.

$$
\begin{array}{r}
\overset{1}{\,}\overset{1}{2}\,6.7\,9 \\
2\,6\,8.1\,0 \\
+\ 6\,9\,0.0\,0 \\
\hline
9\,8\,4.8\,9
\end{array}
$$

Thus, s = $984.89.

<u>Check</u>. We can check by repeating the addition. We can also check by rounding:

26.79 + 268.10 + 690 ≈ 30 + 270 + 690 = 990 ≈ 984.89.

<u>State</u>. $984.89 was spent.

13. <u>Familiarize</u>. We visualize the situation. We let y = the number of years by which the average age in 1984 exceeded the average age in 1961.

19.8	y
23.2	

<u>Translate</u>. This is a "how-much-more" situation.

Average age in 1961	+	Additional years in 1984	is	Average age in 1984
↓	↓	↓	↓	↓
19.8	+	y	=	23.2

<u>Solve</u>. To solve the equation we subtract 19.8 on both sides.

y = 23.2 - 19.8

y = 3.4

$$
\begin{array}{r}
\overset{1}{\cancel{2}}\overset{12}{\cancel{3}}.\overset{12}{\cancel{2}} \\
-\ 1\,9.8 \\
\hline
3.4
\end{array}
$$

<u>Check</u>. We check by adding 3.4 to 19.8 to get 23.2. This checks.

<u>State</u>. The average bride was 3.4 years older in 1984 than in 1961.

15. <u>Familiarize</u>. We visualize the situation. We let
t = the combined time.

4.25 min	4.86 min	3.98 min	5.0 min
t				

<u>Translate</u>. We are combining amounts.

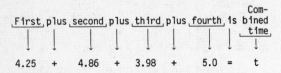

First, plus second, plus third, plus fourth, is Combined time

$$4.25 + 4.86 + 3.98 + 5.0 = t$$

<u>Solve</u>. To solve we carry out the addition.

$$
\begin{array}{r}
\overset{2\ 1}{4.2}5 \\
4.8\ 6 \\
3.9\ 8 \\
+\ 5.0 \\
\hline
1\ 8.0\ 9
\end{array}
$$

Thus, t = 18.09 min.

<u>Check</u>. We can check by repeating the addition.
We can also check by rounding:
4.25 + 4.86 + 3.98 + 5.0 ≈ 4 + 5 + 4 + 5 = 18 ≈
18.09

<u>State</u>. The combined time was 18.09 min.

17. <u>Familiarize</u>. We visualize the situation. We let
D = the distance in miles and K = the distance in
kilometers.

876 mi, or 1401.6 km	1562 mi, or 2499.2 km
D mi or K km		

<u>Translate and Solve</u>. We are combining distances.

Distance from New York to St. Louis	plus	Distance from St. Louis to Los Angeles	is	Distance from New York to Los Angeles

a) Taking distance in miles, we translate to an
equation:

$$876 + 1562 = D$$

To solve we carry out the addition.

$$
\begin{array}{r}
\overset{1\ \ 1}{8}7\ 6 \\
+1\ 5\ 6\ 2 \\
\hline
2\ 4\ 3\ 8
\end{array}
$$

Thus, D = 2438 mi.

b) Taking distance in kilometers, we translate to
an equation:

$$1401.6 + 2499.2 = K$$

To solve we carry out the addition.

17. (continued)

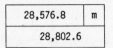

$$
\begin{array}{r}
1\ \overset{1}{4}\overset{1}{0}1.6 \\
+2\ 4\ 9\ 9.2 \\
\hline
3\ 9\ 0\ 0.8
\end{array}
$$

Thus K = 3900.8 km.

<u>Check</u>. We can check by repeating the addition.
We can also check by rounding:
876 + 1562 ≈ 880 + 1560 = 2440 ≈ 2438;

1401.6 + 2499.2 ≈ 1400 + 2500 = 3900 ≈ 3900.8

<u>State</u>. a) It is 2438 mi from New York to Los
Los Angeles.

b) It is 3900.8 km from New York to Los Angeles.

19. <u>Familiarize</u>. We visualize the situation. We let
m = the number of miles driven.

28,576.8	m
28,802.6	

<u>Translate</u>. This is a "how-much-more" situation.

First odometer reading	plus	Miles driven	is	Second odometer reading
28,576.8	+	m	=	28,802.6

<u>Solve</u>. To solve the equation we subtract 28,576.8
on both sides.

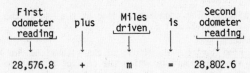

$$m = 28{,}802.6 - 28{,}576.8$$
$$m = 225.8$$

$$
\begin{array}{r}
\overset{\ \ \ \ \ \ \ \ 11}{2\ 8{,}8\overset{7\ 9\ \cancel{\ }\ 16}{\cancel{0}\ \cancel{2}.\cancel{6}}} \\
-\ 2\ 8{,}5\ 7\ 6.8 \\
\hline
2\ 2\ 5.8
\end{array}
$$

<u>Check</u>. We check by adding 225.8 to 28,576.8 to
get 28,802.6. This checks.

<u>State</u>. 225.8 miles were driven.

21. <u>Familiarize</u>. This is a multistep problem. We
will first find the total amount of the checks.
Then we will find how much is left in the account
after the checks are written. Finally, we will
use this amount and the amount of the deposit to
find the balance in the account after all the
changes. We will let c = the total amount of the
checks.

<u>Translate</u>. We are combining amounts.

First check	plus	Second check	plus	Third check	is	Total amount of checks
23.82	+	507.88	+	98.32	=	c

<u>Solve</u>. To solve the equation we carry out the
addition.

$$
\begin{array}{r}
\overset{1\ \ \overset{2}{2}\ \overset{2}{2}\ \overset{1}{1}}{2\ 3.8\ 2} \\
5\ 0\ 7.8\ 8 \\
9\ 8.3\ 2 \\
\hline
6\ 3\ 0.0\ 2
\end{array}
$$

Thus, c = 630.02.

21. (continued)

Now we let a = the amount in the account after the checks are written.

Original amount	less	Check amount	is	New amount
↓	↓	↓	↓	↓
1123.56	−	630.02	=	a

To solve the equation we carry out the subtraction.

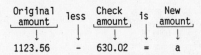

Thus, a = 493.54.

Finally, we let f = the amount in the account after the paycheck is deposited.

Amount after checks	plus	Amount of deposit	is	Final amount
↓	↓	↓	↓	↓
493.54	+	678.20	=	f

We carry out the addition.

$$\overset{1\;1}{4\;9}3.5\,4$$
$$6\;7\;8.2\,0$$
$$1\;1\;7\;1.7\,4$$

Thus, f = 1171.74.

Check. We repeat the calculations.

State. There is $1171.74 in the account after the changes.

23. Familiarize. We visualize the situation. We let p = the number by which the O'Hare passengers exceed the Kennedy passengers, in millions.

45.7	
29.9	p

Translate. We have a "how-much-more" situation.

Kennedy passengers	plus	Additional O'Hare passengers	=	O'Hare passengers
↓	↓	↓	↓	↓
29.9	+	p	=	45.7

Solve. To solve we subtract 29.9 on both sides of the equation.

p = 45.7 − 29.9

p = 15.8

$$\overset{14}{3\,\cancel{4}\,17}$$
$$\cancel{4\;5}.7$$
$$-\;2\;9.9$$
$$1\;5.8$$

Check. We add 15.8 to 29.9 to get 45.7. This checks.

State. O'Hare handles 15.8 million more passengers than Kennedy.

25. Familiarize. We visualize the situation. We let t = the number of passengers Dallas/Ft. Worth and Kennedy handle together, in millions.

32.3	29.9
t	

Translate. We are combining amounts.

Dallas/ Ft. Worth passengers	plus	Kennedy passengers	=	Number handled together
↓	↓	↓	↓	↓
32.3	+	29.9	=	t

Solve. To solve we carry out the addition.

$$\overset{1\;1}{3\;2}.3$$
$$2\;9.9$$
$$6\;2.2$$

Thus, t = 62.2.

Check. To check we can repeat the addition. We can also check by rounding:

32.3 + 29.9 ≈ 30 + 30 = 60 ≈ 62.2

State. Dallas/Ft. Worth and Kennedy handle 62.2 million passengers together.

27. Familiarize. We visualize the situation. We let a = the total amount the student paid.

$39.95	$39.95	$4.80
a		

Translate. We are combining amounts.

First pair of slacks	plus	Second pair of slacks	plus	Tax	is	Total cost
↓	↓	↓	↓	↓	↓	↓
39.95	+	39.95	+	4.80	=	a

Solve. To solve we carry out the addition.

$$\overset{2\;2\;1}{3\;9}.9\,5$$
$$3\;9.9\,5$$
$$+\;\;4.8\,0$$
$$8\;4.7\,0$$

Thus, a = 84.70.

Check. To check we can repeat the addition. We can also check by rounding:

39.95 + 39.95 + 4.80 ≈ 40 + 40 + 5 = 85 ≈ 84.70

State. The student paid $84.70.

29. Familiarize. We let d = the distance around the figure.

Translate. We are combining lengths.

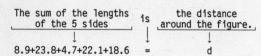

$$8.9+23.8+4.7+22.1+18.6 = d$$

Solve. To solve we carry out the addition.

```
  2 3
    8.9
  2 3.8
    4.7
  2 2.1
+ 1 8.6
  7 8.1
```

Thus, d = 78.1

Check. To check we can repeat the addition. We can also check by rounding:

8.9 + 23.8 + 4.7 + 22.1 + 18.6 ≈

9 + 24 + 5 + 22 + 19 = 79 ≈ 78.1

State. The distance around the figure is 78.1 cm.

31. Familiarize. This is a multistep problem. First we find the sum s of the two 0.8 cm segments. Then we use this length to find d.

Translate.

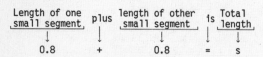

$$0.8 + 0.8 = s$$

Solve. To solve we carry out the addition.

```
  1
  0.8
+ 0.8
  1.6
```

Thus, s = 1.6.

Now we find d.

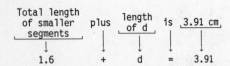

$$1.6 + d = 3.91$$

To solve we subtract 1.6 on both sides of the equation.

d = 3.91 - 1.6

d = 2.31

```
  3.9 1
- 1.6 0
  2.3 1
```

Check. We repeat the calculations.

State. The length d is 2.31 cm.

33. Familiarize. We visualize the situation. We let t = the number of degrees by which the temperature of the bath water exceeds normal body temperature.

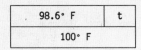

98.6° F	t
100° F	

Translate. We have a "how-much-more" situation.

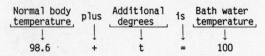

$$98.6 + t = 100$$

Solve. To solve we subtract 98.6 on both sides of the equation.

t = 100 - 98.6

t = 1.4

```
    9 9 10
  1 0 0.0
  -  9 8.6
       1.4
```

Check. To check we add 1.4 to 98.6 to get 100. This checks.

State. The temperature of the bath water is 1.4° F above normal body temperature.

35. Familiarize. This is a multistep problem. We will find the correct change and the amount of the actual change and compare them to determine if the change was correct. We let c = the correct change and a = the actual change.

Translate. We write two equations.

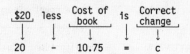

$$20 - 10.75 = c$$

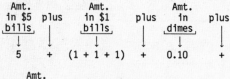

$$5 + (1 + 1 + 1) + 0.10 +$$

Amt. in nickels is Actual change

$$(0.05 + 0.05) = a$$

Solve. To solve the first equation we carry out the subtraction.

```
  1 9 9 10
  2 0.0 0
- 1 0.7 5
    9.2 5
```

Thus, c = 9.25.

To solve the second equation we carry out the addition.

5 + 1 + 1 + 1 + 0.10 + 0.05 + 0.05 = 8.20

Thus, a = 8.20

The correct change is different from the actual change.

35. (continued)

Check. We repeat the calculations.

State. The change was not correct.

37.
$$
\begin{array}{r}
\overset{1}{4}\,\overset{1}{5}\,\overset{1}{6}\,9 \\
+\ 1\ 7\ 6\ 6 \\
\hline
6\ 3\ 3\ 5
\end{array}
$$

39.
$$
\begin{array}{r}
\overset{3}{\cancel{4}}\,\overset{15}{\cancel{5}}\,6\,9 \\
-\ 1\ 7\ 6\ 6 \\
\hline
2\ 8\ 0\ 3
\end{array}
$$

Exercise Set 5.1

1.
```
        8.6     (1 decimal place)
    ×     7     (0 decimal places)
      6 0.2     (1 decimal place)
```

3.
```
        0.8 4   (2 decimal places)
    ×       8   (0 decimal places)
        6.7 2   (2 decimal places)
```

5.
```
        6.3     (1 decimal place)
    × 0.0 4     (2 decimal places)
      0.2 5 2   (3 decimal places)
```

7.
```
        8 7     (0 decimal places)
    × 0.0 0 6   (3 decimal places)
      0.5 2 2   (3 decimal places)
```

9. 10×23.76 23.7.6

1 zero   Move 1 place to the right.

$10 \times 23.76 = 237.6$

11. 1000×783.686852 783.686.852

3 zeros   Move 3 places to the right.

$1000 \times 783.686852 = 783,686.852$

13. 7.8×100 7.80.

2 zeros   Move 2 places to the right.

$7.8 \times 100 = 780$

15. 0.1×89.23 08.9.23

1 decimal place   Move 1 place to the left.

$0.1 \times 89.23 = 8.923$

17. 0.001×97.68 0.097.68

3 decimal places   Move 3 places to the left.

$0.001 \times 97.68 = 0.09768$

19. 78.2×0.01 0.78.2

2 decimal places   Move 2 places to the left.

$78.2 \times 0.01 = 0.782$

21.
```
        3 2.6   (1 decimal place)
    ×     1 6   (0 decimal places)
      1 9 5 6
      3 2 6 0
      5 2 1.6   (1 decimal place)
```

23.
```
        0.9 8 4     (3 decimal places)
    ×       3.3     (1 decimal place)
        2 9 5 2
      2 9 5 2 0
      3.2 4 7 2     (4 decimal places)
```

25.
```
        3 7 4       (0 decimal places)
    ×     2.4       (1 decimal place)
      1 4 9 6
      7 4 8 0
      8 9 7.6       (1 decimal place)
```

27.
```
        7 4 9       (0 decimal places)
    × 0.4 3         (2 decimal places)
      2 2 4 7
      2 9 9 6 0
      3 2 2.0 7     (2 decimal places)
```

29.
```
        0.8 7       (2 decimal places)
    ×     6 4       (0 decimal places)
        3 4 8
      5 2 2 0
      5 5.6 8       (2 decimal places)
```

31.
```
      4 6.5 0       (2 decimal places)
    ×     7 5       (0 decimal places)
      2 3 2 5 0
      3 2 5 5 0 0
      3 4 8 7.5 0   (2 decimal places)
```

Since the last decimal place is 0, we could also write this answer as 3487.5.

33.
```
        8 1.7       (1 decimal place)
    × 0.6 1 2       (3 decimal places)
        1 6 3 4
        8 1 7 0
      4 9 0 2 0 0
      5 0.0 0 0 4   (4 decimal places)
```

35.
```
      1 0.1 0 5         (3 decimal places)
    × 1 1.3 2 4         (3 decimal places)
        4 0 4 2 0
      2 0 2 1 0 0
      3 0 3 1 5 0 0
    1 0 1 0 5 0 0 0
  1 0 1 0 5 0 0 0 0
  1 1 4.4 2 9 0 2 0     (6 decimal places)
```

or 114.42902

37.
```
        0.7 8 9
      ×   1 0 0 0  ←— 3 zeros
```
```
0.789.                1000 × 0.789 = 789
```
Move 3 places
to the right.

39.
```
          1 2.3    (1 decimal place)
        × 1.0 8    (2 decimal places)
          9 8 4
        1 2 3 0 0
        1 3.2 8 4  (3 decimal places)
```

41.
```
          3 2.4    (1 decimal place)
        ×   2.8    (1 decimal place)
        2 5 9 2
        6 4 8
        9 0.7 2    (2 decimal places)
```

43.
```
        0.0 0 3 4 2   (5 decimal places)
      ×       0.8 4   (2 decimal places)
            1 3 6 8
          2 7 3 6 0
      0.0 0 2 8 7 2 8 (7 decimal places)
```

45.
```
          0.3 4 7   (3 decimal places)
        ×   2.0 9   (2 decimal places)
          3 1 2 3
          6 9 4 0 0
        0.7 2 5 2 3 (5 decimal places)
```

47.
```
          3.0 0 5   (3 decimal places)
        × 0.6 2 3   (3 decimal places)
          9 0 1 5
          6 0 1 0 0
        1 8 0 3 0 0 0
        1.8 7 2 1 1 5 (6 decimal places)
```

49.
```
        4 5.6 7 8
      ×     1 0 0 0  ←— 3 zeros
```
```
45.678.      1000 × 45.678 = 45,678
```
Move 3 places
to the right.

51. 100 × 45.678   45.67.8

2 zeros   Move 2 places to the right.

100 × 45.678 = 4567.8

53. Move 2 places to the right

$28.88.¢

Change from $ sign in front to ¢ sign at end

$28.88 = 2888¢

55. Move 2 places to the right

$0.66.¢

Change from $ sign in front to ¢ sign at end

$0.66 = 66¢

57. Move 2 places to the left

$.34.¢

Change from ¢ sign at end to $ sign in front

34¢ = $0.34

59. Move 2 places to the left

$34.45.¢

Change from ¢ sign at end to $ sign in front

3345¢ = $34.45

61. $2.83 trillion = $2.83 × 1,000,000,000,000

                                    12 zeros

$2.830000000000.

Move 12 places to the right

$2.83 trillion = $2,830,000,000,000

63. $2\frac{1}{3} \cdot 4\frac{4}{5} = \frac{7}{3} \cdot \frac{24}{5} = \frac{7 \cdot 3 \cdot 8}{3 \cdot 5}$

$= \frac{3}{3} \cdot \frac{7 \cdot 8}{5} = \frac{56}{5}$

$= 11\frac{1}{5}$

65.
```
            3 4 2
      2 4 ) 8 2 0 8
            7 2 0 0
            1 0 0 8
              9 6 0
                4 8
                4 8
                 0
```

67.   (1 trillion)·(1 billion)

= 1, 000,000,000,000, × 1, 000,000,000,
       12 zeros   9 zeros

= 1, 000,000,000,000,000,000,000,
                  21 zeros

= 10^{21}

Exercise Set 5.2

1.
```
      2.9 9
  2)5.9 8
    4 0 0
    1 9 8
    1 8 0
      1 8
      1 8
         0
```
Divide as though dividing whole numbers. Place the decimal point directly above the decimal point in the dividend.

3.
```
      2 3.7 8
  4)9 5.1 2
    8 0 0 0
    1 5 1 2
    1 2 0 0
      3 1 2
      2 8 0
        3 2
        3 2
           0
```
Divide as though dividing whole numbers. Place the decimal point directly above the decimal point in the dividend.

5.
```
        7.4 8
  1 2)8 9.7 6
      8 4 0 0
        5 7 6
        4 8 0
          9 6
          9 6
             0
```

7.
```
           7.2
  3 3)2 3 7.6
      2 3 1 0
          6 6
          6 6
             0
```

9.
```
      1.1 4 3
  8)9.1 4 4
    8 0 0 0
    1 1 4 4
      8 0 0
      3 4 4
      3 2 0
        2 4
        2 4
           0
```

11.
```
      4.0 4 1
  3)1 2.1 2 3
    1 2 0 0 0
        1 2 3
        1 2 0
            3
            3
            0
```

13.
```
      0.0 7
  5)0.3 5
    3 5
       0
```

15.
```
           7 0.
  0.1 2∧)8.4 0∧
         8 4 0
             0
```
Multiply the divisor by 100 (move the decimal point 2 places). Multiply the same way in the dividend (move 2 places).

Then divide. Place the decimal point in the answer above the new decimal point in the dividend.

17.
```
         2 0.
  3.4∧)6 8.0∧
       6 8 0
           0
           0
           0
```
Put a decimal point at the end of the whole number. Multiply the divisor by 10 (move the decimal point 1 place). Multiply the same way in the dividend (move 1 place), adding an extra 0.

Then divide. Place the decimal point in the answer above the new decimal point in the dividend.

19.
```
       0.4
  1 5)6.0
      6 0
         0
```
Put a decimal point at the end of the whole number.

Write an extra 0 to the right of the decimal point.

Then divide and place the decimal point in the answer above the decimal point in the dividend.

21.
```
        0.4 1
3 6)1 4.7 6
    1 4 4 0
      3 6
      3 6
         0
```

23.
```
         8. 5
3.2∧)2 7.2∧0
    2 5 6
    1 6   0        Write an extra 0
    1 6   0
          0
```

25.
```
          9. 3
4.2∧)3 9.0∧6
    3 7 8 0
    1 2 6
    1 2 6
          0
```

27.
```
    0.6 2 5
8)5.0 0 0
  4 8
    2 0        Write an extra 0
    1 6
      4 0      Write an extra 0
      4 0
         0
```

29.
```
           0. 2 6
0.4 7∧)0.1 2 2 2
       9 4 0
       2 8 2
       2 8 2
             0
```

31.
```
        1 5. 6 2 5
4.8∧)7 5.0∧0 0 0
     4 8 0
     2 7 0
     2 4 0
       3 0 0
       2 8 8
         1 2 0
           9 6
           2 4 0
           2 4 0
                 0
```

33.
```
              2. 3 4
0.0 3 2∧)0.0 7 4∧8 8
         6 4 0 0
         1 0 8 8
           9 6 0
           1 2 8
           1 2 8
                 0
```

35.
```
          0.4 7
8 2)3 8.5 4
    3 2 8 0
      5 7 4
      5 7 4
            0
```

37.
$$\frac{213.4567}{1000}$$

3 zeros

.213.4567
↑__↑
3 places to the left

$$\frac{213.4567}{1000} = 0.2134567$$

39.
$$\frac{213.4567}{10}$$

1 zero

21.3.4567
↑_↑
1 place to the left

$$\frac{213.4567}{10} = 21.34567$$

41.
$$\frac{1.0237}{0.001}$$

3 decimal places

1.023.7
↑__↑
3 places to the right

$$\frac{1.0237}{0.001} = 1023.7$$

43.
$$\frac{56.78}{0.001}$$

3 decimal places

56.780.
↑__↑
3 places to the right

$$\frac{56.78}{0.001} = 56,780$$

45. $4.2 \cdot x = 39.06$

$$\frac{4.2 \cdot x}{4.2} = \frac{39.06}{4.2}$$ Dividing on both sides by
                    by 4.2

$x = 9.3$

```
          9. 3
4.2∧)3 9.0∧6
    3 7 8 0
    1 2 6
    1 2 6
          0
```

47. $1000 \cdot y = 9.0678$

$$\frac{1000 \cdot y}{1000} = \frac{9.0678}{1000}$$ Dividing on both sides by 100

$y = 0.0090678$ Moving the decimal point 3
                places to the left

49.   14 × (82.6 + 67.9)

= 14 × (150.5)   Doing the calculation inside
                        parentheses

= 2107   Multiplying

51.   0.003 + 3.03 ÷ 0.01

= 0.003 + 303   Dividing first

= 303.003   Adding

53.   42 × (10.6 + 0.024)

= 42 × 10.624   Doing the calculation
                        inside parentheses

= 446.208   Multiplying

55.   4.2 × 5.7 + 0.7 ÷ 3.5

= 23.94 + 0.2   Doing the multiplications
                        and divisions in order
                        from left to right

= 24.14   Adding

57.   9.0072 + 0.04 ÷ 0.1²

= 9.0072 + 0.04 ÷ 0.01 Evaluating the
                        exponential expression

= 9.0072 + 4   Dividing

= 13.0072   Adding

59.   (8 − 0.04)² ÷ 4 + 8.7 × 0.4

= (7.96)² ÷ 4 + 8.7 × 0.4   Doing the calculation
                            inside parentheses

= 63.3616 ÷ 4 + 8.7 × 0.4   Evaluating the
                            exponential expression

= 15.8404 + 3.48   Doing the multiplications
                        and divisions in order from
                        left to right

= 19.3204   Adding

61.   86.13 + 95.7 ÷ (9.6 − 0.03)

= 86.13 + 95.7 ÷ 9.57   Doing the calculation
                        inside parentheses

= 86.13 + 10   Dividing

= 96.13   Adding

63.   4 ÷ 0.4 + 0.1 × 5 − 0.1²

= 4 ÷ 0.4 + 0.1 × 5 − 0.01   Evaluating the
                             exponential
                             expression

= 10 + 0.5 − 0.01   Doing the multiplications
                        and divisions in order
                        from left to right

= 10.49   Adding and subtracting in order
                        from left to right

65.   5.5² × [(6 − 4.2) ÷ 0.06 + 0.12]

= 5.5² × [1.8 ÷ 0.06 + 0.12]   Doing the
                               calculation in the
                               innermost
                               parentheses first

= 5.5² × [30 + 0.12]   Doing the
                        calculation inside
                        parentheses

= 5.5² × 30.12

= 30.25 × 30.12   Evaluating the exponential
                        expression

= 911.13   Multiplying

67.   200 × {[(4 − 0.25) ÷ 2.5] − (4.5 − 4.025)}

= 200 × {[3.75 ÷ 2.5] − 0.475}   Doing the
                                 calculations
                                 in the innermost
                                 parentheses first

= 200 × {1.5 − 0.475}   Again, doing the
                        calculations in the
                        innermost parentheses

= 200 × 1.025   Subtracting inside the
                        parentheses

= 205   Multiplying

69. $10\frac{1}{2} + 4\frac{5}{8} = 10\frac{4}{8} + 4\frac{5}{8}$

= $14\frac{9}{8} = 15\frac{1}{8}$

71. $\frac{36}{42} = \frac{6 \cdot 6}{6 \cdot 7} = \frac{6}{6} \cdot \frac{6}{7} = \frac{6}{7}$

Exercise Set 5.3

1. $\frac{3}{5} = \frac{3}{5} \cdot \frac{2}{2}$ We use $\frac{2}{2}$ for 1 to get a
                                                      a denominator of 10.

    = $\frac{6}{10} = 0.6$

3. $\frac{13}{40} = \frac{13}{40} \cdot \frac{25}{25}$ We use $\frac{25}{25}$ for 1 to get a
                                                            denominator of 1000.

    = $\frac{325}{1000} = 0.325$

5. $\frac{1}{5} = \frac{1}{5} \cdot \frac{2}{2} = \frac{2}{10} = 0.2$

7. $\frac{17}{20} = \frac{17}{20} \cdot \frac{5}{5} = \frac{85}{100} = 0.85$

9. $\frac{19}{40} = \frac{19}{40} \cdot \frac{25}{25} = \frac{475}{1000} = 0.475$

11. $\frac{39}{40} = \frac{39}{40} \cdot \frac{25}{25} = \frac{975}{1000} = 0.975$

13. $\frac{13}{25} = \frac{13}{25} \cdot \frac{4}{4} = \frac{52}{100} = 0.52$

15. $\frac{2502}{125} = \frac{2502}{125} \cdot \frac{8}{8} = \frac{20,016}{1000} = 20.016$

17. $\frac{1}{4} = \frac{1}{4} \cdot \frac{25}{25} = \frac{25}{100} = 0.25$

19. $\frac{23}{40} = \frac{23}{40} \cdot \frac{25}{25} = \frac{575}{1000} = 0.575$

21. $\frac{18}{25} = \frac{18}{25} \cdot \frac{4}{4} = \frac{72}{100} = 0.72$

23. $\frac{19}{16} = \frac{19}{16} \cdot \frac{625}{625} = \frac{11,875}{10,000} = 1.1875$

25. $\frac{4}{15} = 4 \div 15$

```
      0.2 6 6
1 5)4.0 0 0
    3 0
    1 0 0
      9 0
      1 0 0
        9 0
        1 0
```

Since 10 keeps reappearing as a remainder, the digits repeat and

$\frac{4}{15} = 0.2666\ldots$ or $0.2\overline{6}$

27. $\frac{1}{3} = 1 \div 3$

```
    .3 3 3
3)1.0 0 0
  9
  1 0
    9
    1 0
      9
      1
```

Since 1 keeps reappearing as a remainder, the digits repeat and

$\frac{1}{3} = 0.333\ldots$ or $0.\overline{3}$

29. $\frac{4}{3} = 4 \div 3$

```
    1.3 3
3)4.0 0
  3
  1 0
    9
    1 0
      9
      1
```

Since 1 keeps reappearing as a remainder, the digits repeat and

$\frac{4}{3} = 1.33\ldots$ or $1.\overline{3}$

31. $\frac{7}{6} = 7 \div 6$

```
    1.1 6 6
6)7.0 0 0
  6
  1 0
    6
    4 0
    3 6
    4 0
    3 6
    4
```

Since 4 keeps reappearing as a remainder, the digits repeat and

$\frac{7}{6} = 1.166\ldots$ or $1.1\overline{6}$

33. $\frac{4}{7} = 4 \div 7$

```
    0.5 7 1 4 2 8 5
7)4.0 0 0 0 0 0 0
  3 5
  5 0
  4 9
  1 0
    7
    3 0
    2 8
    2 0
    1 4
    6 0
    5 6
    4 0
    3 5
    5
```

Since 5 reappears as a remainder, the sequence repeats and

$\frac{4}{7} = 0.571428571428\ldots$ or $0.\overline{571428}$

35. $\frac{11}{12} = 11 \div 12$

```
       0.9 1 6 6
12)1 1.0 0 0 0
    1 0 8
      2 0
      1 2
      8 0
      7 2
      8 0
      7 2
      8 0
```

35. (continued)

Since 8 keeps reappearing as a remainder, the digits repeat and

$\frac{11}{12}$ = 0.91666 . . . or 0.91$\overline{6}$

37. Round 0.2⬚6 6 . . . to the nearest tenth.

       ⌐ Hundredths digit is greater than 5. Round up.

    0.3

Round 0.26⬚6 . . . to the nearest hundredth.

       ⌐ Thousandths digit is greater than 5. Round up.

    0.27

Round 0.266⬚ . . . to the nearest thousandth.

       ⌐ Ten-thousandths digit is greater than 5. Round up.

    0.267

39. Round 0.3⬚3 3 . . . to the nearest tenth.

       ⌐ Hundredths digit is less than 5. Round down.

    0.3

Round 0.33⬚3 . . . to the nearest hundredth.

       ⌐ Thousandths digit is less than 5. Round down.

    0.33

Round 0.333⬚ . . . to the nearest thousandth.

       ⌐ Ten-thousandths digit is less than 5. Round down.

    0.333

41. Round 1.3⬚3 . . . to the nearest tenth.

       ⌐ Hundredths digit is less than 5. Round down.

    1.3

Round 1.33⬚ . . . to the nearest hundredth.

       ⌐ Thousandths digit is less than 5. Round down.

    1.33

Round 1.333⬚ . . . to the nearest thousandth.

       ⌐ Ten-thousandths digit is less than 5. Round down.

    1.333

43. Round 1.1⬚6 6 . . . to the nearest tenth.

       ⌐ Hundredths digit is 5 or higher. Round up.

    1.2

Round 1.16⬚6 . . . to the nearest hundredth.

       ⌐ Thousandths digit is 5 or higher. Round up.

    1.17

Round 1.166⬚ . . . to the nearest thousandth.

       ⌐ Ten-thousandths digit is 5 or higher. Round up.

    1.167

45. 0.$\overline{571428}$

Round to the nearest tenth.

    0.5⬚1428571428 . . .

       ⌐ Hundredths digit is 5 or higher. Round up.

    0.6

Round to the nearest hundredth.

    0.57⬚428571428 . . .

       ⌐ Thousandths digit is less than 5. Round down.

    0.57

Round to the nearest thousandth.

    0.571⬚28571428 . . .

       ⌐ Ten-thousandths digit is less than 5. Round down.

    0.571

47. Round 0.9⬚6 6 . . . to the nearest tenth.

       ⌐ Hundredths digit is less than 5. Round down.

    0.9

Round 0.91⬚6 6 . . . to the nearest hundredth.

       ⌐ Thousandths digit is 5 or higher. Round up.

    0.92

Round 0.916⬚6 . . . to the nearest thousandth.

       ⌐ Ten-thousandths digit is 5 or higher. Round up.

    0.917

49. We will use the first method discussed in the text.

$$\frac{7}{8} \times 12.64 = \frac{7}{8} \times \frac{1264}{100} = \frac{7 \cdot 1264}{8 \cdot 100}$$

$$= \frac{7 \cdot 2 \cdot 2 \cdot 2 \cdot 2 \cdot 79}{2 \cdot 2 \cdot 2 \cdot 2 \cdot 2 \cdot 5 \cdot 5}$$

$$= \frac{2 \cdot 2 \cdot 2 \cdot 2}{2 \cdot 2 \cdot 2 \cdot 2} \cdot \frac{7 \cdot 79}{2 \cdot 5 \cdot 5}$$

$$= 1 \cdot \frac{7 \cdot 79}{2 \cdot 5 \cdot 5}$$

$$= \frac{7 \cdot 79}{2 \cdot 5 \cdot 5} = \frac{553}{50}, \text{ or } 11.06$$

51. We will use the second method discussed in the text.

$$\frac{47}{9} \times 79.95 = 5.\overline{2} \times 79.95 \approx 5.222 \times 79.95 = 417.4989$$

Note that this answer is not as accurate as those found using either of the other methods, due to rounding.

53. We will use the third method discussed in the text.

$$\frac{5}{6} \times 0.0765 + \frac{5}{4} \times 0.1124$$

$$= \frac{5}{6} \times \frac{0.0765}{1} + \frac{5}{4} \times \frac{0.1124}{1}$$

$$= \frac{5 \times 0.0765}{6 \times 1} + \frac{5 \times 0.1124}{4 \times 1}$$

$$= \frac{0.3825}{6} + \frac{0.562}{4}$$

$$= 0.06375 + 0.1405$$

$$= 0.20425$$

55. $9 \cdot 2\frac{1}{3} = \frac{9}{1} \cdot \frac{7}{3} = \frac{9 \cdot 7}{3} = \frac{3 \cdot 3 \cdot 7}{3 \cdot 1} = \frac{3}{3} \cdot \frac{3 \cdot 7}{1} = 21$

57.
$$\begin{array}{r} 20 = 19\frac{5}{5} \\ - 16\frac{3}{5} = -16\frac{3}{5} \\ \hline 3\frac{2}{5} \end{array}$$

59. a) Find the prime factorization of each number.
$25 = 5 \cdot 5, \quad 65 = 5 \cdot 13$

   b) Create a product by writing factors, using each the greatest number of times it occurs in any one factorization.
   The LCM is $5 \cdot 5 \cdot 13$, or 325.

61. Use a calculator to find $1 \div 7$. The result, on a calculator with a 10 digit readout, is 0.142857142. The digits in the first six decimal places will continue to repeat.

$$\frac{1}{7} = 0.\overline{142857}$$

63. Use a calculator to find $3 \div 7$. The result, on a calculator with a 10 digit readout, is 0.428571428. The digits in the first six decimal places will continue to repeat.

$$\frac{3}{7} = 0.\overline{428571}$$

65. Use a calculator to find $5 \div 7$. The result, on a calculator with a 10 digit readout, is 0.714285714. The digits in the first six decimal places will continue to repeat.

$$\frac{5}{7} = 0.\overline{714285}$$

67. Use a calculator to find $1 \div 9$. The result, on a calculator with a 10 digit readout, is 0.111111111. The digit 1 repeats to the right of the decimal point.

$$\frac{1}{9} = 0.\overline{1}$$

69. Use a calculator to find $1 \div 999$. The result, on a calculator with a 10 digit readout, is 0.001001001. The digits in the first three decimal places will continue to repeat.

$$\frac{1}{999} = 0.\overline{001}$$

71. Use a calculator to find $1 \div 81$. The result, on a calculator with a 10 digit readout, is 0.012345679. We need a longer readout to determine if this is a repeating decimal. We can confirm that it is by doing the division by hand.

$$\frac{1}{81} = 0.\overline{012345679}$$

73. Use a calculator to find $100 \div 81$. The result, on a calculator with a 10 digit readout, is 1.234567901. The digits to the right of the decimal will continue to repeat.

$$\frac{100}{81} = 1.\overline{234567901}$$

75. We find decimal notation for each fraction, writing the first six decimal places.

$\frac{2}{3} = 0.666666, \quad \frac{15}{19} = 0.789473, \quad \frac{11}{13} = 0.846153,$

$\frac{5}{7} = 0.714285, \quad \frac{13}{15} = 0.866666, \quad \frac{17}{20} = 0.850000$

We arrange the fractions in order from smallest to largest by comparing digits in the decimal notation as discussed on page 183.

$$\frac{2}{3}, \frac{5}{7}, \frac{15}{19}, \frac{11}{13}, \frac{17}{20}, \frac{13}{15}$$

Exercise Set 5.4

1. We are estimating the sum
   $109.95 + $249.95.

   We round $109.95 to the nearest ten and $249.95 to the nearest ten. The estimate is
   $110 + $250 = $360.

   Answer (d) is correct.

<u>3</u>. We are estimating the difference

$299 - $249.95.

We round $299 to the nearest ten and $249.95 to the nearest ten. The estimate is

$300 - $250 = $50.

Answer (c) is correct.

<u>5</u>. We are estimating the product

9 × $299.

We round $299 to the nearest ten. The estimate is

9 × $300 = $2700.

Answer (a) is correct.

<u>7</u>. We are estimating the quotient

$1700 ÷ $299.

Rounding $299, we get $300. Since $1700 is close to $1800, which is a multiple of $300, we estimate

$1800 ÷ $300, so the answer is about 6.

Answer (c) is correct.

<u>9</u>. This is about 0.0 + 1.3 + 0.3, so the answer is about 1.6.

<u>11</u>. This is about 6 + 0 + 0, so the answer is about 6.

<u>13</u>. This is about 52 + 1 + 7, so the answer is about 60.

<u>15</u>. This is about 2.7 - 0.4, so the answer is about 2.3.

<u>17</u>. This is about 200 - 20, so the answer is about 180.

<u>19</u>. This is about 50 × 8, rounding 49 to the nearest ten and 7.89 to the nearest one, so the answer is about 400. Answer (a) is correct.

<u>21</u>. This is about 100 × 0.08, rounding 98.4 to the nearest ten and 0.083 to the nearest hundredth, so the answer is about 8. Answer (c) is correct.

<u>23</u>. This is about 4 ÷ 4, so the answer is about 1. Answer (b) is correct.

<u>25</u>. This is about 75 ÷ 25, so the answer is about 3. Answer (b) is correct.

<u>27</u>.

$$
\begin{array}{r}
3 \\
3\overline{)9} \\
3\overline{)27} \\
2\overline{)54} \\
2\overline{)108}
\end{array}
$$

108 = 2·2·3·3·3

<u>29</u>. $\dfrac{125}{400} = \dfrac{25 \cdot 5}{25 \cdot 16} = \dfrac{25}{25} \cdot \dfrac{5}{16} = \dfrac{5}{16}$

<u>31</u>. We round each factor to the nearest ten. The estimate is 180 × 60 = 10,800. The estimate is close to the result given, so the decimal point was placed correctly.

Exercise Set 5.5

<u>1</u>. <u>Familiarize</u>. Repeated addition fits this situation. We let c = the cost of 7 blouses.

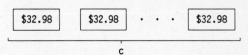

<u>Translate</u>.

Price per blouse	times	Number of blouses	is	Total cost
↓	↓	↓	↓	↓
32.98	×	7	=	c

<u>Solve</u>. To solve the equation we carry out the multiplication.

$$
\begin{array}{r}
3\,2.9\,8 \\
\times \qquad 7 \\
\hline
2\,3\,0.8\,6
\end{array}
$$

Thus, c = 230.86.

<u>Check</u>. We obtain a partial check by rounding and estimating:

32.98 × 7 ≈ 30 × 7 = 210 ≈ 230.86

<u>State</u>. Seven blouses cost $230.86.

<u>3</u>. <u>Familiarize</u>. Repeated addition fits this situation. We let c = the cost of 17.7 gal of gasoline.

<u>Translate</u>.

Cost per gallon	times	Number of gallons	is	Total cost
↓	↓	↓	↓	↓
1.699	·	17.7	=	c

<u>Solve</u>. To solve the equation we carry out the multiplication.

$$
\begin{array}{r}
1.6\,9\,9 \\
\times \quad 1\,7.7 \\
\hline
1\,1\,8\,9\,3 \\
1\,1\,8\,9\,3\,0 \\
1\,6\,9\,9\,0\,0 \\
\hline
3\,0.0\,7\,2\,3
\end{array}
$$

Thus, c = 30.0723.

<u>Check</u>. We obtain a partial check by rounding and estimating:

1.699 × 17.7 ≈ 1.5 × 20 = 30 ≈ 30.0723

<u>State</u>. We round $30.0723 to the nearest cent and get the cost to be about $30.07.

5. <u>Familiarize</u>. Think of this as a rectangular array with 250 objects (miles) arranged in 4 rows. We want to find how many objects (miles) are in each row. We let m represent this number.

<u>Translate</u>. We think (Total miles) ÷ (Number of hours) = (Number of miles in 1 hour).

$$250 ÷ 4 = m$$

<u>Solve</u>. To solve the equation we carry out the division.

```
      6 2.5
4)2 5 0.0
  2 4 0
    1 0
      8
      2 0
      2 0
         0
```

Thus, m = 62.5.

<u>Check</u>. We obtain a partial check by rounding and estimating:

$$250 ÷ 4 ≈ 300 ÷ 5 = 60 ≈ 62.5$$

<u>State</u>. The car went 62.5 mi in 1 hour.

7. <u>Familiarize</u>. Think of a rectangular array containing $47.60 arranged in 8 rows. We want to find how many dollars are in each row. We let s represent this number.

<u>Translate</u>. We think (Total cost) ÷ (Number of students) = (Each student's share).

$$47.60 ÷ 8 = s$$

<u>Solve</u>. To solve the equation we carry out the division.

```
      5.9 5
8)4 7.6 0
  4 0 0 0
    7 6 0
    7 2 0
      4 0
      4 0
         0
```

Thus, s = 5.95.

<u>Check</u>. We obtain a partial check by rounding and estimating:

$$47.60 ÷ 8 ≈ 50 ÷ 10 = 5 ≈ 5.95$$

<u>State</u>. Each person's share is $5.95.

9. <u>Familiarize</u>. We draw a picture, letting A = the area.

```
┌──────────────────┐
│        A         │   312.6 ft
└──────────────────┘
  800.4 ft
```

<u>Translate</u>. We use the formula A = ℓ·w.

$$A = 800.4 × 312.6$$

9. (continued)

<u>Solve</u>. We carry out the multiplication.

```
      3 1 2.6
    × 8 0 0.4
    1 2 5 0 4
2 5 0 0 8 0 0 0
2 5 0,2 0 5.0 4
```

Thus, A = 250,205.04.

<u>Check</u>. We obtain a partial check by rounding and estimating:

$$800.4 × 312.6 ≈ 800 × 300 = 240,000 ≈ 250,205.04$$

<u>State</u>. The area is 250,205.04 sq ft.

11. <u>Familiarize</u>. Repeated addition fits this situation. We let c = the cost of 6 albums.

<u>Translate</u>.

Price per album	times	Number of albums	is	Total cost
↓	↓	↓	↓	↓
8.88	×	6	=	c

<u>Solve</u>. To solve the equation we carry out the multiplication.

```
    8.8 8
  ×     6
  5 3.2 8
```

Thus, c = 53.28.

<u>Check</u>. We obtain a partial check by rounding and estimating:

$$8.88 × 6 ≈ 9 × 6 = 54 ≈ 53.28$$

<u>State</u>. Six albums cost $53.28.

13. <u>Familiarize</u>. Repeated addition fits this situation. We let t = the total savings in 1 year.

<u>Translate</u>.

Number of weeks	times	Savings per week	is	Total savings
↓	↓	↓	↓	↓
52	·	2.68	=	t

<u>Solve</u>. To solve the equation we carry out the multiplication.

```
    2.6 8
  × 5 2
    5 3 6
1 3 4 0 0
1 3 9.3 6
```

Thus, t = 139.36.

13. (continued)

Check. We obtain a partial check by rounding and estimating:

$52 \times 2.68 \approx 50 \times 3 = 150 \approx 139.36$

State. In 1 year, $139.36 would be saved.

15. Familiarize. Think of a rectangular array containing $11,178.72 arranged in 24 rows. We want to find how many dollars are in each row. We let p represent this number.

Translate. We think (Amount of loan) ÷ (Number of payments) = (Amount of each payment).

$11,178.72 \div 24 = p$

Solve. To solve the equation we carry out the division.

```
          4 6 5.7 8
   2 4 ) 1 1,1 7 8.7 2
          9 6 0 0 0 0
          1 5 7 8 7 2
          1 4 4 0 0 0
            1 3 8 7 2
            1 2 0 0 0
              1 8 7 2
              1 6 8 0
                1 9 2
                1 9 2
                    0
```

Thus, p = 465.78.

Check. We obtain a partial check by rounding and estimating:

$11,178.72 \div 24 \approx 10,000 \div 20 = 500 \approx \$465.78$

State. Each payment is $465.78.

17. Familiarize. Repeated addition fits this situation. We let d = the distance the plane travels.

Translate.

Number of km flown in 1 hr	times	Number of hours flown	is	Total distance
↓	↓	↓	↓	↓
147.9	×	6	=	d

Solve. To solve the equation we carry out the multiplication.

```
  1 4 7.9
  ×     6
  8 8 7.4
```

Thus, d = 887.4.

Check. We obtain a partial check by rounding and estimating:

$147.9 \times 6 \approx 150 \times 6 = 900 \approx 887.4$

State. The plane travels 887.4 km.

19. Familiarize. This is a two-step problem. First, we find the mileage cost of driving 120 miles at 27¢ per mile. We let c represent this number.

Translate and Solve.

Cost per mile	times	Number of miles	is	Mileage cost
↓	↓	↓	↓	↓
0.27	·	120	=	c

To solve the equation we carry out the multiplication.

```
    1 2 0
  × 0.2 7
    8 4 0
  2 4 0 0
  3 2.4 0
```

Thus, c = 32.40.

Next, we add the rental cost for one day to the mileage cost to find the total cost. We let y represent this number.

Cost for one day	plus	Mileage cost	is	Total cost for one day
↓	↓	↓	↓	↓
24.95	+	32.40	=	y

To solve the equation we carry out the addition.

```
    2 4.9 5
  + 3 2.4 0
    5 7.3 5
```

Thus, y = 57.35.

Check. To check, we first subtract the daily charge from the total cost to get the total mileage cost:

$57.35 - 24.95 = 32.40$

Then we divide the total mileage cost by the cost per mile:

$32.4 \div 27 = 120$

The number 57.35 checks.

State. The total cost of driving 120 miles in 1 day is $57.35.

21. <u>Familiarize</u>. This is a two-step problem. First, we find the number of miles that have been driven between fillups. This is a "how-much-more" situation. We let n = the number of miles driven.

<u>Translate and Solve</u>.

First odometer reading	plus	Number of miles driven	is	Second odometer reading
↓	↓	↓	↓	↓
26,342.8	+	n	=	26,736.7

To solve the equation we subtract 26,342.8 on both sides

n = 26,736.7 - 26,342.8   2 6,7 3 6.7

n = 393.9   - 2 6,3 4 2.8

                                        3 9 3.9

Second, we divide the total number of miles driven by the number of gallons. This gives us m = the number of miles per gallon.

393.9 ÷ 19.5 = m

To find the number m, we divide.

```
              2 0. 2
1 9.5 ∧ ⟌3 9 3 3.9 ∧ 0
        3 9 0 0
          3 9   0
          3 9   0
              0
```

Thus, m = 20.2.

<u>Check</u>. To check, we first multiply the number of miles per gallon times the number of gallons.

19.5 × 20.2 = 393.9

Then we add 393.9 to 26,342.8:

26,342.8 + 393.9 = 26,736.7

The number 20.2 checks.

<u>State</u>. The driver gets 20.2 miles per gallon.

23. <u>Familiarize</u>. We visualize a rectangular array consisting of 748.45 objects with 62.5 objects in each row. We want to find n, the number of rows.

<u>Translate</u>. We think (Total number of pounds) ÷ (Pounds per cubic foot) = (Number of cubic feet).

748.45 ÷ 62.5 = n

<u>Solve</u>. We carry out the division.

```
                1 1. 9 7 5 2
6 2.5 ∧ ⟌7 4 8.4 ∧ 5 0 0 0
        6 2 5 0 0
        1 2 3 4 5
          6 2 5 0
          6 0 9 5
          5 6 2 5
            4 7 0 0
            4 3 7 5
            3 2 5 0
            3 1 2 5
            1 2 5 0
            1 2 5 0
                  0
```

Thus, n = 11.9752.

<u>Check</u>. We obtain a partial check by rounding and estimating:

748.45 ÷ 62.5 ≈ 700 ÷ 70 = 10 ≈ 11.9752

<u>State</u>. The tank holds 11.9752 cubic feet of water.

25. <u>Familiarize</u>. This is a two-step problem. First, we find the number of games that can be played in one hour. Think of an array containing 60 minutes (1 hour = 60 minutes) with 1.5 minutes in each row. We want to find how many rows there are. We let g represent the number.

<u>Translate and Solve</u>. We think (Number of minutes) ÷ (Number of minutes per game) = (Number of games).

60 ÷ 1.5 = g

To solve the equation we carry out the division.

```
            4 0.
1.5 ∧ ⟌6 0.0 ∧
      6 0 0
          0
          0
          0
```

Thus, g = 40. Second, we find the cost t of playing 40 video games. Repeated addition fits this situation. (We express 25¢ as $0.25.)

Cost of one game	times	Number of games played	is	Total cost
↓	↓	↓	↓	↓
0.25	×	40	=	t

To solve the equation we carry out the multiplication.

25. (continued)

$$
\begin{array}{r}
0.2\,5 \\
\times\ \ \ 4\,0 \\
\hline
1\,0.0\,0
\end{array}
$$

Thus, t = 10.

Check. To check, we first divide the total cost by the cost per game to find the number of games played:

10 ÷ 0.25 = 40

Then we multiply 40 by 1.5 to find the total time:

1.5 × 40 = 60

The number 10 checks.

State. It costs $10 to play video games for one hour.

27. Familiarize. This is a two-step problem. First, we find how many minutes there are in 2 hr. We let m represent this number. Repeated addition fits this situation. (Remember that 1 hr = 60 min.)

Translate and Solve.

Number of minutes in 1 hour	times	Number of hours	is	Total number of minutes
↓	↓	↓	↓	↓
60	·	2	=	m

To solve the equation we carry out the multiplication.

$$
\begin{array}{r}
6\,0 \\
\times\ \ \ 2 \\
\hline
1\,2\,0
\end{array}
$$

Thus, m = 120.

Next, we find how many calories are burned in 120 minutes. We let t represent this number. Repeated addition fits this situation also.

Number of calories burned in 1 minute	times	Number of minutes	is	Total number of calories burned
↓	↓	↓	↓	↓
8.6	×	120	=	t

To solve the equation we carry out the multiplication.

$$
\begin{array}{r}
1\,2\,0 \\
\times\ \ \ 8.6 \\
\hline
7\,2\,0 \\
9\,6\,0\,0 \\
\hline
1\,0\,3\,2.0
\end{array}
$$

Thus, t = 1032.

27. (continued)

Check. To check, we first divide the total calories by the number of calories burned in one minute to find the total number of minutes the person mowed:

1032 ÷ 8.6 = 120

Then we divide 120 by 60 to find the number of hours:

120 ÷ 60 = 2

The number 1032 checks.

State. In 2 hr of mowing, 1032 calories would be burned.

29. Familiarize. This is a two-step problem. First we find the number of miles that are driven between fillups. We let n represent this number.

Translate and Solve.

First odometer reading	plus	Number of miles driven	is	Second odometer reading
↓	↓	↓	↓	↓
36,057.1	+	n	=	36,217.6

To solve the equation we subtract 36,057.1 on both sides.

n = 36,217.6 - 36,057.1

n = 160.5

$$
\begin{array}{r}
3\,6,2\,1\,7.6 \\
-\ 3\,6,0\,5\,7.1 \\
\hline
1\,6\,0.5
\end{array}
$$

Then we divide the number of miles driven by the number of gallons required to fill the tank at 36,217.6 mi. This gives us m = the number of miles per gallon.

$$
\begin{array}{r}
1\,4\,.4\,5 \\
1\,1.1_\wedge\overline{)1\,6\,0.5_\wedge 0\,0} \\
\underline{1\,1\,1\,0} \\
4\,9\,5 \\
\underline{4\,4\,4} \\
5\,1\ \ 0 \\
\underline{4\,4\ \ 4} \\
6\ \ 6\,0 \\
\underline{5\ \ 5\,5} \\
1\ \ 0\,5
\end{array}
$$

We stop dividing at this point, because we will round to the nearest tenth. We find m ≈ 14.5.

Check. To check, we first multiply the number of miles per gallon times the number of gallons to find the number of miles driven between fillups:

14.5 × 11.1 = 160.95

Then we add 160.95 to 36,057.1:

36,057.1 + 160.95 = 36,218.05

This is slightly different from the final odometer reading because we rounded, but it is close enough that we can be reasonably certain that 14.5 checks.

State. The driver gets about 14.5 miles per gallon.

31. <u>Familiarize</u>. The batting average is the number of hits per times at bat. We let a = the batting average.

<u>Translate</u>. We think (Number of hits) ÷ (Number of at bats) = (Batting average).

$$187 \div 610 = a$$

<u>Solve</u>. We carry out the division.

```
        0.3 0 6 5
6 1 0 )1 8 7.0 0 0 0
        1 8 3 0
          4 0 0 0
          3 6 6 0
            3 4 0 0
            3 0 5 0
              3 5 0
```

We stop dividing at this point, because we will round to the nearest thousandth. We find a ≈ 0.307.

<u>Check</u>. We obtain a partial check by rounding and estimating:

$$187 \div 610 \approx 200 \div 600 = 0.\overline{3} \approx 0.307$$

<u>State</u>. Jose Canseco's batting average was about 0.307.

33. <u>Familiarize</u>. We let d = the amount earned each day. (We will assume the student earned the same amount each day.)

<u>Translate</u>. We think (Total earnings) ÷ (Number of days) = (Amount earned each day).

$$78.27 \div 9 = d$$

<u>Solve</u>. We carry out the division.

```
      8.6 9 6
9 )7 8.2 7 0
    7 2 0 0
      6 2 7
      5 4 0
        8 7
        8 1
          6 0
          5 4
            6
```

Thus, d = 8.69$\overline{6}$.

<u>Check</u>. We obtain a partial check by rounding and estimating

$$78.27 \div 9 \approx 80 \div 10 = 8 \approx 8.69\overline{6}$$

<u>State</u>. We round 8.69$\overline{6}$ and get the amount earned each day to be about $8.70.

35. <u>Familiarize</u>. We let c = the cost per kilogram.

<u>Translate</u>. We think (Total cost) ÷ (Total kilograms) = (Cost per kilogram).

$$16.58 \div 4 = c$$

<u>Solve</u>. We carry out the division.

```
      4.1 4 5
4 )1 6.5 8 0
   1 6 0 0
      5 8
      4 0
      1 8
      1 6
        2 0
        2 0
          0
```

Thus, c = 4.145.

<u>Check</u>. We obtain a partial check by rounding and estimating:

$$16.58 \div 4 \approx 16 \div 4 = 4 \approx 4.145$$

<u>State</u>. We round 4.145 and get the cost per kilogram to be about $4.15.

37. <u>Familiarize</u>. We let c = the cost per acre.

<u>Translate</u>. We think (Total cost) ÷ (Number of acres) = (Cost per acre).

$$18,716.47 \div 47.5 = c$$

<u>Solve</u>. We carry out the division.

```
              3 9 4 .0 3 0
4 7.5 )1 8,7 1 6.4 7 0 0
        1 4 2 5 0 0 0
          4 4 6 6 4 7
          4 2 7 5 0 0
            1 9 1 4 7
            1 9 0 0 0
              1 4 7 0
              1 4 2 5
                4 5 0
```

We stop dividing at this point, because we will round to the nearest cent (or hundredth). We find c ≈ 394.03.

<u>Check</u>. We obtain a partial check by rounding and estimating:

$$18,716.47 \div 47.5 \approx 20,000 \div 50 = 400 \approx 394.03$$

<u>State</u>. The cost per acre was about $394.03.

39. <u>Familiarize</u>. We make a drawing. We shade the area left over after the pool is built.

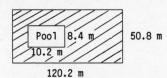

Pool 8.4 m   50.8 m
10.2 m
120.2 m

This is a three-step problem. We will find the area A of the lot and the area S of the swimming pool. Then we will find R, the area left over after the pool is built.

<u>Translate and Solve</u>.

A = ℓ·w = 120.2 × 50.8 = 6106.16
S = ℓ·w = 10.2 × 8.4 = 85.68

Area left over	is	Area of lot	minus	Area of pool
↓	↓	↓	↓	↓
R	=	6106.16	-	85.68

We carry out the subtraction.

```
  6 1 0 6.1 6
-     8 5.6 8
  6 0 2 0.4 8
```

Thus, R = 6020.48.

<u>Check</u>. We obtain a partial check by rounding and estimating:

A = 120.2 × 50.8 ≈ 120 × 50 = 6000 ≈ 6106.16
S = 10.2 × 8.4 ≈ 10 × 8 = 80 ≈ 85.68
R ≈ 6000 - 80 = 5920 ≈ 6020

<u>State</u>. The area left over is 6020.48 sq m.

41. <u>Familiarize</u>. This is a three-step problem. First we find m = the number of months in 30 years.

<u>Translate and Solve</u>.

Number of years	times	Number of months in a year	is	Total number of months
↓	↓	↓	↓	↓
30	·	12	=	m

We carry out the multiplication.

```
    1 2
  × 3 0
    3 6 0
```

Thus, m = 360.

Next we find a = the amount paid back.

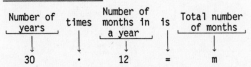

Monthly payment	times	Number of months	is	Amount paid back
↓	↓	↓	↓	↓
880.52	·	360	=	a

41. (continued)

We carry out the multiplication.

```
      8 8 0.5 2
    ×     3 6 0
    5 2 8 3 1 2 0
  2 6 4 1 5 6 0 0
  3 1 6,9 8 7.2 0
```

Finally, we find p = the amount by which the amount paid back exceeds the amount of the loan. This is a "how-much-more" situation.

Amount of loan	plus	Excess amount	is	Amount paid back
↓	↓	↓	↓	↓
120,000	+	p	=	316,987.20

We subtract 120,000 on both sides.

```
p = 316,987.20 - 120,000        3 1 6,9 8 7.2 0
p = 196,987.20                 - 1 2 0,0 0 0.0 0
                                 1 9 6,9 8 7.2 0
```

<u>Check</u>. We repeat the calculations.

<u>State</u>. You pay back $196,987.20 more than the amount of the loan.

43. $24\frac{1}{4} = 24\frac{3}{12} = 23\frac{15}{12}$

   $-10\frac{2}{3} = -10\frac{8}{12} = -10\frac{8}{12}$

   $\phantom{-10\frac{2}{3} = -10\frac{8}{12} = -} 13\frac{7}{12}$

45. $24\frac{1}{4} = 24\frac{3}{12}$

   $+10\frac{2}{3} = +10\frac{8}{12}$

   $\phantom{+10\frac{2}{3} = +} 34\frac{11}{12}$

47. a) The cost of the car wash is $3.39. We compute:

   $0.999(10) + $3.39 = $13.38

   b) The cost of the car wash is $4.50. We compute:

   $0.999(10) + $4.50 = $14.49

   c) 13.38 < 14.49, so method (a) is cheaper.

   d) The cost of the car wash is $3.39, (103.9 cents = $1.039). We compute:

   $1.039(10) + $3.39 = $13.78

   e) 13.78 < 14.49, so method (d) is cheaper.

Exercise Set 6.1

1. The ratio of 4 to 5 can be written $\frac{4}{5}$.

3. The ratio of 0.4 to 12 can be written $\frac{0.4}{12}$.

5. The ratio of milk to flour can be written $\frac{2}{12}$.

7. The ratio of $2.7 billion to $13.1 billion can be written $\frac{2.7}{13.1}$.

   The ratio of $13.1 billion to $2.7 billion can be written $\frac{13.1}{2.7}$.

9. We can use cross products:

   $5 \cdot 9 = 45$ $6 \cdot 7 = 42$

   Since the cross products are not the same, we know that the numbers are not proportional.

11. We can use cross products:

   $1 \cdot 20 = 20$ $2 \cdot 10 = 20$

   Since the cross products are the same, we know that the numbers are proportional.

13. $\frac{18}{4} = \frac{x}{10}$

   $18 \cdot 10 = 4 \cdot x$ Finding cross products

   $\frac{18 \cdot 10}{4} = x$ Dividing by 4

   $\frac{180}{4} = x$ Multiplying

   $45 = x$ Dividing

15. $\frac{x}{8} = \frac{9}{6}$

   $6 \cdot x = 8 \cdot 9$ Finding cross products

   $x = \frac{8 \cdot 9}{6}$ Dividing by 6

   $x = \frac{72}{6}$ Multiplying

   $x = 12$ Dividing

17. $\frac{t}{12} = \frac{5}{6}$

   $6 \cdot t = 12 \cdot 5$

   $t = \frac{12 \cdot 5}{6}$

   $t = \frac{60}{6}$

   $t = 10$

19. $\frac{2}{5} = \frac{8}{n}$

   $2 \cdot n = 5 \cdot 8$

   $n = \frac{5 \cdot 8}{2}$

   $n = \frac{40}{2}$

   $n = 20$

21. $\frac{n}{15} = \frac{10}{30}$

   $30 \cdot n = 15 \cdot 10$

   $n = \frac{15 \cdot 10}{30}$

   $n = \frac{150}{30}$

   $n = 5$

23. $\frac{16}{12} = \frac{24}{x}$

   $16 \cdot x = 12 \cdot 24$

   $x = \frac{12 \cdot 24}{16}$

   $x = \frac{288}{16}$

   $x = 18$

25. $\frac{6}{11} = \frac{12}{x}$

   $6 \cdot x = 11 \cdot 12$

   $x = \frac{11 \cdot 12}{6}$

   $x = \frac{132}{6}$

   $x = 22$

27. $\frac{20}{7} = \frac{80}{x}$

   $20 \cdot x = 7 \cdot 80$

   $x = \frac{7 \cdot 80}{20}$

   $x = \frac{560}{20}$

   $x = 28$

29. $\frac{12}{9} = \frac{x}{7}$

   $12 \cdot 7 = 9 \cdot x$

   $\frac{12 \cdot 7}{9} = x$

   $\frac{84}{9} = x$

   $\frac{28}{3} = x$ Simplifying

   $9\frac{1}{3} = x$ Writing a mixed numeral

31. $\frac{x}{13} = \frac{2}{9}$

    $9 \cdot x = 13 \cdot 2$

    $x = \frac{13 \cdot 2}{9}$

    $x = \frac{26}{9}$ or $2\frac{8}{9}$

33. $\frac{t}{0.16} = \frac{0.15}{0.40}$

    $0.40 \times t = 0.16 \times 0.15$

    $t = \frac{0.16 \times 0.15}{0.40}$

    $t = \frac{0.024}{0.40}$

    $t = 0.06$

35. $\frac{25}{100} = \frac{n}{20}$

    $25 \cdot 20 = 100 \cdot n$

    $\frac{25 \cdot 20}{100} = n$

    $\frac{500}{100} = n$

    $5 = n$

37. $\frac{7}{\frac{1}{4}} = \frac{28}{x}$

    $7 \cdot x = 28 \cdot \frac{1}{4}$

    $7 \cdot x = 7$ Multiplying on the right side

    $x = \frac{7}{7}$

    $x = 1$

39. $\frac{\frac{1}{4}}{\frac{1}{2}} = \frac{\frac{1}{2}}{x}$

    $\frac{1}{4} \cdot x = \frac{1}{2} \cdot \frac{1}{2}$

    $\frac{1}{4} \cdot x = \frac{1}{4}$ Multiplying on the right side

    $x = \frac{1}{4} \div \frac{1}{4}$

    $x = \frac{1}{4} \cdot 4$ Multiplying by the reciprocal

    $x = 1$

41. $\frac{1}{2} = \frac{7}{x}$

    $1 \cdot x = 2 \cdot 7$

    $x = \frac{2 \cdot 7}{1}$

    $x = \frac{14}{1}$

    $x = 14$

43. $\frac{\frac{2}{7}}{\frac{3}{4}} = \frac{\frac{5}{6}}{y}$

    $\frac{2}{7} \cdot y = \frac{3}{4} \cdot \frac{5}{6}$

    $\frac{2}{7} \cdot y = \frac{15}{24}$

    $y = \frac{15}{24} \div \frac{2}{7}$

    $y = \frac{15}{24} \cdot \frac{7}{2}$

    $y = \frac{15 \cdot 7}{24 \cdot 2}$

    $y = \frac{3 \cdot 5 \cdot 7}{3 \cdot 8 \cdot 2}$

    $y = \frac{3}{3} \cdot \frac{5 \cdot 7}{8 \cdot 2}$

    $y = \frac{35}{16}$, or $2\frac{3}{16}$

45. We find the cross products:

    $12 \cdot 4 = 48$ $8 \cdot 6 = 48$

    Since the cross products are equal,

    $\frac{12}{8} = \frac{6}{4}$.

47.
    ```
         5 0
      4)2 0 0
        2 0 0
            0
            0
            0
    ```

49.
    ```
           1 4.5
      1 6)2 3 2.0
          1 6 0
            7 2
            6 4
              8 0
              8 0
               0
    ```

51. Use a calculator.

    The ratio of people to sheep is

    $\frac{13,339,000}{145,304,000}$ or about 0.09.

    The ratio of sheep to people is

    $\frac{145,304,000}{13,339,000}$ or about 10.89.

Exercise Set 6.2

1. $\frac{120 \text{ km}}{3 \text{ hr}}$, or $40 \frac{\text{km}}{\text{hr}}$

3. $\frac{440 \text{ m}}{40 \text{ sec}}$, or $11 \frac{\text{m}}{\text{sec}}$

5. $\frac{342 \text{ yd}}{2.25 \text{ days}}$, or $152 \frac{\text{yd}}{\text{day}}$

$$
\begin{array}{r}
152. \\
2.25_\wedge\overline{)3\,4\,2.0\,0_\wedge} \\
2\,2\,5\,0\,0 \\
\overline{1\,1\,7\,0\,0} \\
1\,1\,2\,5\,0 \\
\overline{4\,5\,0} \\
4\,5\,0 \\
\overline{0}
\end{array}
$$

7. $\frac{500 \text{ km}}{20 \text{ hr}} = 25 \frac{\text{km}}{\text{hr}}$

   $\frac{20 \text{ hr}}{500 \text{ km}} = 0.04 \frac{\text{hr}}{\text{km}}$

9. $\frac{623 \text{ gal}}{1000 \text{ sq ft}} = 0.623 \frac{\text{gal}}{\text{sq ft}}$

11. $\frac{30 \text{ servings}}{12 \text{ lb}} = 2.5 \frac{\text{servings}}{\text{lb}}$

13. $\frac{186,000 \text{ mi}}{1 \text{ sec}} = 186,000 \frac{\text{mi}}{\text{sec}}$

15. $\frac{4.6 \text{ km}}{2 \text{ hr}} = 2.3 \frac{\text{km}}{\text{hr}}$

17. $\frac{2660 \text{ mi}}{4.75 \text{ hr}} = 560 \frac{\text{mi}}{\text{hr}}$

19. Unit price $= \dfrac{\text{Price}}{\text{Number of units}} = \dfrac{\$33.25}{3.5 \text{ yd}}$

    $= \$9.50/\text{yd}$

21. Unit price $= \dfrac{\text{Price}}{\text{Number of units}} = \dfrac{\$2.79}{13 \text{ oz}} = \dfrac{279\cent}{13 \text{ oz}}$

    $\approx 21.46 \frac{\cent}{\text{oz}}$

23. $\dfrac{\$1.35}{1\frac{1}{4} \text{ lb}} = \dfrac{\$1.35}{1.25 \text{ lb}} = \$1.08/\text{lb}$

25. Compare the unit prices.

    For Brand A: $\dfrac{\$1.79}{18 \text{ oz}} \approx \$0.099/\text{oz}$

    For Brand B: $\dfrac{\$1.65}{16 \text{ oz}} \approx \$0.103/\text{oz}$

    Thus, Brand A has the lower unit price.

27. Compare the unit prices. Recall that 1 qt = 32 oz, so 1 qt, 14 oz = 32 oz + 14 oz = 46 oz.

    For Brand A: $\dfrac{\$1.29}{46 \text{ oz}} = \dfrac{129\cent}{46 \text{ oz}} \approx 2.80 \frac{\cent}{\text{oz}}$

    For Brand B: $\dfrac{49\cent}{18 \text{ oz}} \approx 2.72 \frac{\cent}{\text{oz}}$

    Thus, Brand B has the lower unit price.

29. Compare the unit prices.

    For Brand A: $\dfrac{\$1.00}{3 \text{ bars}} \approx \$0.33/\text{bar}$

    For Brand B: $\dfrac{\$1.29}{4 \text{ bars}} \approx \$0.32/\text{bar}$

    Thus, Brand B has the lower unit price.

31. Compare the unit prices. Recall that 1 lb = 16 oz, so 1 lb, 8 oz = 16 oz + 8 oz = 24 oz.

    For Brand A: $\dfrac{\$1.29}{7 \text{ oz}} \approx \$0.184/\text{oz}$

    For Brand B: $\dfrac{\$3.96}{24 \text{ oz}} = \$0.165 \text{ oz}$

    Thus, Brand B has the lower unit price.

33. First, find the total number of ounces in each carton. The carton with six 16-oz bottles has

    $6 \times 16 = 96$ oz.

    The carton with eight 16-oz bottles has

    $8 \times 16 = 128$ oz.

    Now compare the unit prices.

    For the six bottle carton: $\dfrac{\$4.49}{96 \text{ oz}} \approx \$0.047/\text{oz}$

    For the eight bottle carton: $\dfrac{\$5.49}{128 \text{ oz}} \approx \$0.043/\text{oz}$

    The carton with eight 16-oz bottles has the lower unit price.

35. **Familiarize.** We visualize the situation. We let p = the number by which the number of piano players exceeds the number of guitar players, in millions.

18.9 million	p
20.6 million	

    Translate. This is a "how-many-more" situation.

Number of guitar players	+	Additional number of piano players	=	Number of piano players
↓	↓	↓	↓	↓
18.9	+	p	=	20.6

35. (continued)

Solve. To solve the equation we subtract 18.9 on both sides.

p = 20.6 - 18.9

p = 1.7

$$\begin{array}{r} \overset{1\ 9\ \ 16}{2\,0.6} \\ -\ 1\ 8.9 \\ \hline 1.7 \end{array}$$

Check. We repeat the calculation.

State. There are 1.7 million more piano players than guitar players.

37. $\underline{100} \times 678.19$ 678.19.

2 zeros   Move 2 places to the right

$100 \times 678.19 = 67{,}819$

39. Use a calculator to divide.

$$\frac{500 \text{ m}}{43.33 \text{ sec}} \approx 11.54 \frac{\text{m}}{\text{sec}}$$

$$\frac{43.33 \text{ sec}}{500 \text{ m}} = 0.08666 \frac{\text{sec}}{\text{m}}$$

Exercise Set 6.3

1. Let d represent the distance traveled in 42 days.

$$\begin{array}{l} \text{Distance} \longrightarrow \\ \text{Time} \longrightarrow \end{array} \frac{234}{14} = \frac{d}{42} \begin{array}{l} \longleftarrow \text{Distance} \\ \longleftarrow \text{Time} \end{array}$$

Solve: $234 \cdot 42 = 14 \cdot d$ Finding cross products

$\dfrac{234 \cdot 42}{14} = d$ Dividing by 14

$\dfrac{234 \cdot 3 \cdot 14}{14} = d$ Factoring

$234 \cdot 3 = d$ Simplifying

$702 = d$ Dividing

The car would travel 702 km in 42 days.

3. Let x represent the cost of 9 sweatshirts.

$$\begin{array}{l} \text{Sweatshirts} \longrightarrow \\ \text{Dollars} \longrightarrow \end{array} \frac{2}{18.80} = \frac{9}{x} \begin{array}{l} \longleftarrow \text{Sweatshirts} \\ \longleftarrow \text{Dollars} \end{array}$$

Solve: $2 \cdot x = 18.80 \cdot 9$ Finding cross products

$x = \dfrac{18.80 \cdot 9}{2}$ Dividing by 2

$x = \dfrac{2 \cdot 9.40 \cdot 9}{2}$ Factoring

$x = 9.40 \cdot 9$ Simplifying

$x = 84.6$ Dividing

Thus, 9 sweatshirts cost $84.60.

5. Let A represent the earned run average.

$$\begin{array}{l} \text{Earned runs} \longrightarrow \\ \text{Innings} \longrightarrow \end{array} \frac{A}{9} = \frac{71}{179} \begin{array}{l} \longleftarrow \text{Earned runs} \\ \longleftarrow \text{Innings} \end{array}$$

Solve: $179 \cdot A = 9 \cdot 71$

$A = \dfrac{9 \cdot 71}{179}$

$A = \dfrac{639}{179}$ Multiplying

$A \approx 3.57$ Dividing and rounding to the nearest hundredth

Tom Browning's earned run average was about 3.57.

7. Let D represent the number of deer in the game preserve.

$$\begin{array}{l} \text{Deer tagged} \longrightarrow \\ \text{originally} \\ \text{Deer in game} \longrightarrow \\ \text{preserve} \end{array} \frac{318}{D} = \frac{56}{168} \begin{array}{l} \longleftarrow \text{Tagged deer} \\ \longleftarrow \text{caught later} \\ \longleftarrow \text{Deer caught later} \end{array}$$

Solve: $318 \cdot 168 = 56 \cdot D$

$\dfrac{318 \cdot 168}{56} = D$

$\dfrac{2 \cdot 3 \cdot 53 \cdot 2 \cdot 2 \cdot 2 \cdot 3 \cdot 7}{2 \cdot 2 \cdot 2 \cdot 7} = D$

$2 \cdot 3 \cdot 53 \cdot 3 = D$

$954 = D$

We estimate that there are 954 deer in the game preserve.

9. Let x represent the number of pounds needed for 54 servings.

$$\begin{array}{l} \text{Pounds} \longrightarrow \\ \text{Servings} \longrightarrow \end{array} \frac{8}{36} = \frac{x}{54} \begin{array}{l} \longleftarrow \text{Pounds} \\ \longleftarrow \text{Servings} \end{array}$$

Solve: $8 \cdot 54 = 36 \cdot x$

$\dfrac{8 \cdot 54}{36} = x$

$\dfrac{2 \cdot 2 \cdot 2 \cdot 2 \cdot 3 \cdot 3 \cdot 3}{2 \cdot 2 \cdot 3 \cdot 3} = x$

$2 \cdot 2 \cdot 3 = x$

$12 = x$

Thus, 12 pounds are needed for 54 servings.

11. Let t represent the number of trees required to produce 391 pounds of coffee.

$$\text{Trees} \longrightarrow \quad \frac{14}{17} = \frac{t}{391} \quad \longleftarrow \text{Trees}$$
$$\text{Pounds} \longrightarrow \qquad\qquad \longleftarrow \text{Pounds}$$

Solve: $14 \cdot 391 = 17 \cdot t$

$$\frac{14 \cdot 391}{17} = t$$

$$\frac{14 \cdot 17 \cdot 23}{17} = t$$

$$14 \cdot 23 = t$$

$$322 = t$$

Thus, 322 trees are required to produce 391 pounds of coffee.

13. Let z represent the number of pounds of zinc in the alloy.

$$\text{Zinc} \longrightarrow \quad \frac{3}{13} = \frac{z}{520} \quad \longleftarrow \text{Zinc}$$
$$\text{Copper} \longrightarrow \qquad\qquad \longleftarrow \text{Copper}$$

Solve: $3 \cdot 520 = 13 \cdot z$

$$\frac{3 \cdot 520}{13} = z$$

$$\frac{3 \cdot 13 \cdot 2 \cdot 2 \cdot 2 \cdot 5}{13} = z$$

$$3 \cdot 2 \cdot 2 \cdot 2 \cdot 5 = z$$

$$120 = z$$

There are 120 lb of zinc in the alloy.

15. Let d represent the number of defective bulbs in a lot of 22,000.

$$\begin{array}{l}\text{Defective} \\ \text{bulbs} \end{array} \longrightarrow \quad \frac{18}{200} = \frac{d}{22{,}000} \quad \longleftarrow \begin{array}{l}\text{Defective} \\ \text{bulbs} \end{array}$$
$$\text{Bulbs in lot} \longrightarrow \qquad\qquad \longleftarrow \text{Bulbs in lot}$$

Solve: $18 \cdot 22{,}000 = 200 \cdot d$

$$\frac{18 \cdot 22{,}000}{200} = d$$

$$\frac{18 \cdot 110 \cdot 200}{200} = d$$

$$18 \cdot 110 = d$$

$$1980 = d$$

There would be 1980 defective bulbs in a lot of 22,000.

17. Let d represent the actual distance between the cities.

$$\text{Map distance} \longrightarrow \quad \frac{1}{16.6} = \frac{3.5}{d} \quad \longleftarrow \text{Map distance}$$
$$\text{Actual distance} \longrightarrow \qquad\qquad \longleftarrow \text{Actual distance}$$

Solve: $1 \cdot d = 16.6 \cdot 3.5$

$$d = 58.1$$

The cities are 58.1 mi apart.

19. Let w represent the number of inches of water to which $5\frac{1}{2}$ ft of snow will melt.

$$\text{Snow} \longrightarrow \quad \frac{1\frac{1}{2}}{2} = \frac{5\frac{1}{2}}{w} \quad \longleftarrow \text{Snow}$$
$$\text{Water} \longrightarrow \qquad\qquad \longleftarrow \text{Water}$$

Solve: $1\frac{1}{2} \cdot w = 2 \cdot 5\frac{1}{2}$

$$\frac{3}{2} \cdot w = \frac{2}{1} \cdot \frac{11}{2} \qquad \begin{array}{l}\text{Writing fractional} \\ \text{notation} \end{array}$$

$$\frac{3}{2} \cdot w = \frac{2 \cdot 11}{1 \cdot 2}$$

$$\frac{3}{2} \cdot w = 11 \qquad \text{Simplifying}$$

$$w = 11 \div \frac{3}{2}$$

$$w = \frac{11}{1} \cdot \frac{2}{3}$$

$$w = \frac{11 \cdot 2}{1 \cdot 3}$$

$$w = \frac{22}{3}, \text{ or } 7\frac{1}{3}$$

Thus, $5\frac{1}{2}$ ft of snow will melt to $7\frac{1}{3}$ in. of water.

21. Let s represent the amount the student will spend during the academic year if spending continues at the given rate. Note that the length of the academic year is 16 weeks + 16 weeks = 32 weeks.

$$\text{Weeks} \longrightarrow \quad \frac{3}{50} = \frac{32}{s} \quad \longleftarrow \text{Weeks}$$
$$\text{Spending} \longrightarrow \qquad\qquad \longleftarrow \text{Spending}$$

Solve: $3 \cdot s = 50 \cdot 32$

$$s = \frac{50 \cdot 32}{3}$$

$$s = \frac{1600}{3}$$

$$s \approx 533.33$$

At the given rate, the student would spend about $533.33 during the academic year. Thus, the budget will not be adequate.

To find when the money will be exhausted, we let w represent the number of weeks it will take to spend $400 if spending continues at the given rate.

$$\text{Weeks} \longrightarrow \quad \frac{3}{50} = \frac{w}{400} \quad \longleftarrow \text{Weeks}$$
$$\text{Spending} \longrightarrow \qquad\qquad \longleftarrow \text{Spending}$$

Solve: $3 \cdot 400 = 50 \cdot w$

$$\frac{3 \cdot 400}{50} = w$$

$$\frac{3 \cdot 2 \cdot 2 \cdot 2 \cdot 5 \cdot 5}{2 \cdot 5 \cdot 5} = w$$

$$3 \cdot 2 \cdot 2 \cdot 2 = w$$

$$24 = w$$

21. (continued)

The money will be gone in 24 weeks.

To find how much more will be needed to complete the year we subtract the amount available from the amount needed.

$533.33 - $400 = $133.33

The student will need $133.33 to complete the year.

23. Let R represent the earned runs.

Earned runs $\longrightarrow$ $\dfrac{2.63}{9} = \dfrac{R}{7356}$ $\longleftarrow$ Earned runs
Innings $\longrightarrow$ $\phantom{\dfrac{2.63}{9}}$ $\longleftarrow$ Innings

Solve: $2.63 \cdot 7356 = 9 \cdot R$

$\dfrac{2.63 \cdot 7356}{9} = R$

$\dfrac{19{,}346.28}{9} = R$

$2150 \approx R$

Cy Young gave up about 2150 earned runs.

Exercise Set 7.1

1. $90\% = \dfrac{90}{100}$ A ratio of 90 to 100

   $90\% = 90 \times \dfrac{1}{100}$ Replacing % by $\times \dfrac{1}{100}$

   $90\% = 90 \times 0.01$ Replacing % by $\times 0.01$

3. $12.5\% = \dfrac{12.5}{100}$ A ratio of 12.5 to 100

   $12.5\% = 12.5 \times \dfrac{1}{100}$ Replacing % by $\times \dfrac{1}{100}$

   $12.5\% = 12.5 \times 0.01$ Replacing % by $\times 0.01$

5. 67%   67   .67.   67% = 0.67

            Drop   Move decimal
            percent   point 2 places
            symbol   to the left

7. 45.6%   45.6   0.45.6   45.6% = 0.456

            Drop   Move decimal
            percent   point 2 places
            symbol   to the left

9. 59.01%   59.01   .59.01   59.01% = 0.5901

            Drop   Move decimal
            percent   point 2 places
            symbol   to the left

11. 10%   10   .10.   10% = 0.10, or 0.1

            Drop   Move decimal
            percent   point 2 places
            symbol   to the left

13. 1%   1   0.01.   1% = 0.01

            Drop   Move decimal
            percent   point 2 places
            symbol   to the left

15. 200%   2.00.   200% = 2

17. 0.1%   0.00.1   0.1% = 0.001

19. 0.09%   0.00.09   0.09% = 0.0009

21. 0.18%   0.18   0.00.18   0.18% = 0.0018

            Drop   Move decimal
            percent   point 2 places
            symbol   to the left

23. 23.19%   23.19   0.23.19   23.19% = 0.2319

            Drop   Move decimal
            percent   point 2 places
            symbol   to the left

25. 90%   0.90.   90% = 0.9

27. 10.8%   .10.8   10.8% = 0.108

29. 45.8%   0.45.8   45.8% = 0.458

31. 0.47   0.47.   47%   0.47 = 47%

            Move decimal   Write
            point 2 places   a %
            to the right   symbol

33. 0.03   0.03.   3%   0.03 = 3%

            Move decimal   Write
            point 2 places   a %
            to the right   symbol

35. 1.00   1.00.   100%   1.00 = 100%

            Move decimal   Write
            point 2 places   a %
            to the right   symbol

37. 0.334   0.33.4   33.4%   0.334 = 33.4%

            Move decimal   Write
            point 2 places   a %
            to the right   symbol

39. 0.75.   75%   0.75 = 75%

41. 0.40.   40%   0.4 = 40%

43. 0.00.6   0.6%   0.006 = 0.6%

45. 0.01.7   1.7%   0.017 = 1.7%

47. 0.27.18   27.18%   0.2718 = 27.18%

49. 0.02.39   2.39%   0.0239 = 2.39%

51. 0.02.5   2.5%   0.025 = 2.5%

53. 0.24.   24%   0.24 = 24%

55. To convert $\frac{100}{3}$ to a mixed numeral, we divide.

$$
\begin{array}{r}
3\ 3 \\
3\overline{)1\ 0\ 0} \\
\underline{9\ 0} \\
1\ 0 \\
\underline{9} \\
1
\end{array}
$$

$\frac{100}{3} = 33\ \frac{1}{3}$

57. To convert $\frac{2}{3}$ to decimal notation, we divide.

$$
\begin{array}{r}
.6\ 6 \\
3\overline{)2.0\ 0} \\
\underline{1\ 8} \\
2\ 0 \\
\underline{1\ 8} \\
2
\end{array}
$$

Since 2 keeps reappearing as a remainder, the digits repeat and

$\frac{2}{3} = 0.66\ldots$ or $0.\overline{6}$.

59. Multiply by 100. (This is equivalent to moving the decimal point two places to the right.)

Exercise Set 7.2

1. We use the definition of percent.

$\frac{41}{100} = 41\%$

3. We use the definition of percent.

$\frac{1}{100} = 1\%$

5. We multiply by 1 to get 100 in the denominator.

$\frac{2}{10} = \frac{2}{10} \cdot \frac{10}{10} = \frac{20}{100} = 20\%$

7. We multiply by 1 to get 100 in the denominator.

$\frac{3}{10} = \frac{3}{10} \cdot \frac{10}{10} = \frac{30}{100} = 30\%$

9. $\frac{1}{2} = \frac{1}{2} \cdot \frac{50}{50} = \frac{50}{100} = 50\%$

11. Find decimal notation by division.

$$
\begin{array}{r}
0.6\ 2\ 5 \\
8\overline{)5.0\ 0\ 0} \\
\underline{4\ 8} \\
2\ 0 \\
\underline{1\ 6} \\
4\ 0 \\
\underline{4\ 0} \\
0
\end{array}
$$

Convert to percent notation.

0.62.5 $\frac{5}{8} = 62.5\%$ or $62\frac{1}{2}\%$

13. $\frac{2}{5} = \frac{2}{5} \cdot \frac{20}{20} = \frac{40}{100} = 40\%$

15. Find decimal notation by division.

$$
\begin{array}{r}
0.6\ 6\ 6 \\
3\overline{)2.0\ 0\ 0} \\
\underline{1\ 8} \\
2\ 0 \\
\underline{1\ 8} \\
2\ 0 \\
\underline{1\ 8} \\
2
\end{array}
$$

We get a repeating decimal: $0.06\overline{6}$

Convert to percent notation.

0.66.$\overline{6}$ $\frac{2}{3} = 66.\overline{6}\%$ or $66\frac{2}{3}\%$

17. $$
\begin{array}{r}
0.1\ 6\ 6 \\
6\overline{)1.0\ 0\ 0} \\
\underline{6} \\
4\ 0 \\
\underline{3\ 6} \\
4\ 0 \\
\underline{3\ 6} \\
4
\end{array}
$$

We get a repeating decimal: $0.16\overline{6}$

0.16.$\overline{6}$ $\frac{1}{6} = 16.\overline{6}\%$ or $16\frac{2}{3}\%$

19. $\frac{4}{25} = \frac{4}{25} \cdot \frac{4}{4} = \frac{16}{100} = 16\%$

21. $\frac{1}{20} = \frac{1}{20} \cdot \frac{5}{5} = \frac{5}{100} = 5\%$

23. $\frac{17}{50} = \frac{17}{50} \cdot \frac{2}{2} = \frac{34}{100} = 34\%$

25. $\frac{9}{25} = \frac{9}{25} \cdot \frac{4}{4} = \frac{36}{100} = 36\%$

27. $80\% = \dfrac{80}{100}$ Definition of percent

$= \dfrac{4 \cdot 20}{5 \cdot 20}$ ⎫

$= \dfrac{4}{5} \cdot \dfrac{20}{20}$ ⎬ Simplifying

$= \dfrac{4}{5}$ ⎭

29. $62.5\% = \dfrac{62.5}{100}$ Definition of percent

$= \dfrac{62.5}{100} \cdot \dfrac{10}{10}$ Multiplying by 1 to eliminate the decimal point in the numerator

$= \dfrac{625}{1000}$

$= \dfrac{5 \cdot 125}{8 \cdot 125}$ ⎫

$= \dfrac{5}{8} \cdot \dfrac{125}{125}$ ⎬ Simplifying

$= \dfrac{5}{8}$ ⎭

31. $33\frac{1}{3}\% = \dfrac{100}{3}\%$ Converting from mixed numeral to fractional notation

$= \dfrac{100}{3} \times \dfrac{1}{100}$ Definition of percent

$= \dfrac{100 \cdot 1}{3 \cdot 100}$ Multiplying

$= \dfrac{1}{3} \cdot \dfrac{100}{100}$ ⎫

$= \dfrac{1}{3}$ ⎬ Simplifying

33. $16.\overline{6}\% = 16\frac{2}{3}\%$ $\left(16.\overline{6} = 16\frac{2}{3}\right)$

$= \dfrac{50}{3}\%$ Converting from mixed numeral to fractional notation

$= \dfrac{50}{3} \times \dfrac{1}{100}$ Definition of percent

$= \dfrac{50 \cdot 1}{3 \cdot 50 \cdot 2}$ Multiplying

$= \dfrac{1}{6} \cdot \dfrac{50}{50}$ ⎫

$= \dfrac{1}{6}$ ⎬ Simplifying

35. $7.25\% = \dfrac{7.25}{100} = \dfrac{7.25}{100} \cdot \dfrac{100}{100}$

$= \dfrac{725}{10,000} = \dfrac{29}{400} \cdot \dfrac{25}{25} = \dfrac{29}{400}$

37. $0.8\% = \dfrac{0.8}{100} = \dfrac{0.8}{100} \times \dfrac{10}{10} = \dfrac{8}{1000} = \dfrac{1}{125} \cdot \dfrac{8}{8} = \dfrac{1}{125}$

39. $35\% = \dfrac{35}{100} = \dfrac{7}{20} \cdot \dfrac{5}{5} = \dfrac{7}{20}$

41. $\dfrac{1}{8} = 1 \div 8$

$$\begin{array}{r} .1\,2\,5 \\ 8\,\overline{)1.0\,0\,0} \\ \underline{8} \\ 2\,0 \\ \underline{1\,6} \\ 4\,0 \\ \underline{4\,0} \\ 0 \end{array}$$

$\dfrac{1}{8} = 0.125 = 12\frac{1}{2}\%$ or 12.5%

$\dfrac{1}{6} = 1 \div 6$

$$\begin{array}{r} .1\,6\,6 \\ 6\,\overline{)1.0\,0\,0} \\ \underline{6} \\ 4\,0 \\ \underline{3\,6} \\ 4\,0 \\ \underline{3\,6} \\ 4 \end{array}$$ We get a repeating decimal: $0.1\overline{6}$

$0.16.\overline{6}$ $0.1\overline{6} = 16.\overline{6}\%$

$\dfrac{1}{6} = 0.1\overline{6} = 16.\overline{6}\%$ or $16\frac{2}{3}\%$

$20\% = \dfrac{20}{100} = \dfrac{1}{5} \cdot \dfrac{20}{20} = \dfrac{1}{5}$

$.20.$

$\dfrac{1}{5} = 0.2 = 20\%$

$0.25.$ $0.25 = 25\%$

$25\% = \dfrac{25}{100} = \dfrac{1}{4} \cdot \dfrac{25}{25} = \dfrac{1}{4}$

$\dfrac{1}{4} = 0.25 = 25\%$

$33\frac{1}{3}\% = \dfrac{100}{3}\% = \dfrac{100}{3} \times \dfrac{1}{100} = \dfrac{100}{300} = \dfrac{1}{3} \cdot \dfrac{100}{100} = \dfrac{1}{3}$

$.33.\overline{3}$ $33.\overline{3}\% = 0.33\overline{3}$ or $0.\overline{3}$

$\dfrac{1}{3} = 0.\overline{3} = 33\frac{1}{3}\%$ or $33.\overline{3}\%$

$37.5\% = \dfrac{37.5}{100} = \dfrac{37.5}{100} \cdot \dfrac{10}{10} = \dfrac{375}{1000}$

$= \dfrac{3}{8} \cdot \dfrac{125}{125} = \dfrac{3}{8}$

$.37.5$ $37.5\% = 0.375$

41. (continued)

$\frac{3}{8} = 0.375 = 37\frac{1}{2}\%$ or 37.5%

$40\% = \frac{40}{100} = \frac{2}{5} \cdot \frac{20}{20} = \frac{2}{5}$

.40. $40\% = 0.4$

$\frac{2}{5} = 0.4 = 40\%$

$\frac{3}{5} = \frac{3}{5} \cdot \frac{20}{20} = \frac{60}{100} = 60\%$

.60. $60\% = 0.6$

$\frac{3}{5} = 0.6 = 60\%$

0.62.5 $0.625 = 62.5\%$ or $62\frac{1}{2}\%$

$62.5\% = \frac{62.5}{100} = \frac{62.5}{100} \cdot \frac{10}{10} = \frac{625}{1000} = \frac{5}{8} \cdot \frac{125}{125} = \frac{5}{8}$

$\frac{5}{8} = 0.625 = 62.5\%$

$\frac{2}{3} = 2 \div 3$

$\begin{array}{r} .6\,6 \\ 3\overline{)2.0\,0} \\ \underline{1\,8} \\ 2\,0 \\ \underline{1\,8} \\ 2 \end{array}$ We get a repeating decimal: $0.\overline{6}$

0.66.$\overline{6}$ $0.\overline{6} = 66.\overline{6}\%$ or $66\frac{2}{3}\%$

$\frac{2}{3} = 0.\overline{6} = 66.\overline{6}\%$ or $66\frac{2}{3}\%$

$75\% = \frac{75}{100} = \frac{3}{4} \cdot \frac{25}{25} = \frac{3}{4}$

$\frac{3}{4} = 0.75 = 75\%$

$\frac{4}{5} = \frac{4}{5} \cdot \frac{20}{20} = \frac{80}{100} = 80\%$

.80. $80\% = 0.8$

$\frac{4}{5} = 0.8 = 80\%$

.83.$\overline{3}$ $83.\overline{3}\% = 0.83\overline{3}$ or $0.8\overline{3}$

41. (continued)

$\frac{5}{6} = 0.8\overline{3} = 83\frac{1}{3}\%$ or $83.\overline{3}\%$

.87.5 $87.5\% = 0.875$

$\frac{7}{8} = 0.875 = 87\frac{1}{2}\%$ or 87.5%

$100\% = \frac{100}{100} = \frac{1}{1} = 1$

$\frac{1}{1} = 1 = 100\%$

43. $15 \cdot y = 75$

$\frac{15 \cdot y}{15} = \frac{75}{15}$

$y = 5$

$\begin{array}{r} 5 \\ 1\,5\overline{)7\,5} \\ \underline{7\,5} \\ 0 \end{array}$

45. $3 = 0.16 \times b$

$\frac{3}{0.16} = \frac{0.16 \times b}{0.16}$

$18.75 = b$

$\begin{array}{r} 1\,8.\,7\,5 \\ 0.1\,6_\wedge\overline{)3.0\,0_\wedge 0\,0} \\ \underline{1\,6\,0} \\ 1\,4\,0 \\ \underline{1\,2\,8} \\ 1\,2\ \,0 \\ \underline{1\,1\ \,2} \\ 8\,0 \\ \underline{8\,0} \\ 0 \end{array}$

47. Use a calculator.

$\frac{54}{999} = 0.05.405405\ldots = 5.\overline{405}\%$

Exercise Set 7.3

1. What is 41% of 89?

$\downarrow \quad \downarrow \ \downarrow \quad \downarrow \quad \downarrow$

a   = 41% × 89

3. 89 is what percent of 99?

$\downarrow \ \downarrow \qquad \downarrow \qquad \downarrow \ \downarrow$

89 =   n   × 99

5. 13 is 25% of what?

$\downarrow \ \downarrow \ \downarrow \quad \downarrow \quad \downarrow$

13 = 25% ×   b

7. What is 120% of 75?
↓   ↓   ↓   ↓   ↓
a = 120% × 75   Translating

Convert 120% to decimal notation and multiply.

```
    7 5
  × 1.2          (120% = 1.2)
  1 5 0
  7 5 0
  9 0.0
```

120% of 75 is 90.

9. 150% of 30 is what?
↓   ↓   ↓   ↓   ↓
150% × 30 = a   Translating

Convert 150% to decimal notation and multiply.

```
    3 0
  × 1.5          (150% = 1.5)
  1 5 0
  3 0 0
  4 5.0
```

150% of 30 is 45.

11. What is 5% of $300?
↓   ↓   ↓   ↓   ↓
a = 5% × 300   Translating

Convert 5% to decimal notation and multiply.

```
    3 0 0
  × 0.0 5         (5% = 0.05)
  1 5.0 0
```

$15 is 5% of $300.

13. 2.1% of 50 is what?
↓   ↓   ↓   ↓   ↓
2.1% × 50 = a   Translating

Convert 2.1% to decimal notation and multiply.

```
      5 0
  × .0 2 1        (2.1% = 0.021)
      5 0
  1 0 0 0
  1.0 5 0
```

2.1% of 50 is 1.05.

15. $12 is what percent of $50?
↓   ↓   ↓   ↓   ↓
12 = n × 50   Translating

To solve the equation we divide on both sides by 50 and convert the answer to percent notation.

$12 \div 50 = n$

$\frac{12}{50} = n$

$\frac{12}{50} \cdot \frac{2}{2} = n$

$\frac{24}{100} = n$

24% = n

$12 is 24% of $50.

17. 20 is what percent of 10?
↓   ↓   ↓   ↓   ↓
20 = n × 10   Translating

To solve the equation we divide on both sides by 10 and convert the answer to percent notation.

20 ÷ 10 = n

2 = n

200% = n

20 is 200% of 10.

19. What percent of $300 is $150?
↓   ↓   ↓   ↓   ↓
n × 300 = 150   Translating

n × 300 = 150

n = 150 ÷ 300

n = 0.5

n = 50%

50% of $300 is $150.

21. What percent of 80 is 100?
↓   ↓   ↓   ↓   ↓
n × 80 = 100   Translating

n × 80 = 100

n = 100 ÷ 80

n = 1.25

n = 125%

125% of 80 is 100.

23. 20 is 50% of what?
$$\downarrow \quad \downarrow \quad \downarrow \quad \downarrow \quad \downarrow$$
20 = 50% ×   b   Translating

To solve the equation we divide on both sides by 50%:

20 ÷ 50% = b

20 ÷ 0.5 = b   (50% = 0.5)

40 = b

```
              4 0.
0.5∧ √2 0.0∧
              2 0 0
                  0
                  0
                  0
```

20 is 50% of 40.

25. 40% of what is $16?
$$\downarrow \quad \downarrow \quad \downarrow \quad \downarrow \quad \downarrow$$
40% ×   b   = 16   Translating

To solve the equation we divide on both sides by 40%:

b = 16 ÷ 40%

b = 16 ÷ 0.4   (40% = 0.4)

b = 40

```
              4 0
0.4∧ √1 6.0∧
              1 6 0
                  0
                  0
                  0
```

40% of $40 is $16.

27. 56.32 is 64% of what?
$$\downarrow \quad \downarrow \quad \downarrow \quad \downarrow \quad \downarrow$$
56.32 = 64% ×   b   Translating

56.32 ÷ 64% = b

56.32 ÷ 0.64 = b

88 = b

```
                  8 8
0.6 4∧ √5 6.3 2∧
                  5 1 2 0
                    5 1 2
                    5 1 2
                        0
```

56.32 is 64% of 88.

29. 70% of what is 14?
$$\downarrow \quad \downarrow \quad \downarrow \quad \downarrow \quad \downarrow$$
70% ×   b   = 14   Translating

b = 14 ÷ 70%

b = 14 ÷ 0.7

b = 20

```
                  2 0
0.7∧ √1 4.0∧
                  1 4 0
                      0
                      0
                      0
```

70% of 20 is 14.

31. What is $62\frac{1}{2}$% of 10?
$$\downarrow \quad \downarrow \quad \downarrow \quad \downarrow \quad \downarrow$$
a   = $62\frac{1}{2}$% × 10   Translating

a = 0.625 × 10 $\left[62\frac{1}{2}\% = 0.625\right]$

a = 6.25   Multiplying

6.25 is $62\frac{1}{2}$% of 10.

33. What is 8.3% of $10,200?
$$\downarrow \quad \downarrow \quad \downarrow \quad \downarrow \quad \downarrow$$
a   = 8.3% × 10,200

a = 0.083 × 10,200   (8.3% = 0.083)

a = 846.6

$846.60 is 8.3% of $10,200.

35. $\underline{0.09}$ = $\frac{9}{100}$

2 decimal
places   2 zeros

37. $\frac{89}{100}$ .89.
                           ⌐⌐
2 zeros   Move 2 places

$\frac{89}{100}$ = 0.89

39. Estimate: Round 7.75% to 8% and $10,880 to
$11,000. Then translate:

What is 8% of $11,000?
$$\downarrow \quad \downarrow \quad \downarrow \quad \downarrow \quad \downarrow$$
a   = 8% × 11,000

This tells us what to do.

```
  1 1,0 0 0
×      0.0 8          (8% = 0.08)
  8 8 0.0 0
```

$880 is about 7.75% of $10,880. (Answers may vary.)

39. (continued)

Calculate: First we translate.

What is 7.75% of $10,880?

$$a = 7.75\% \times 10,880$$

Use a calculator to multiply:

0.0775 × 10,880 = 843.2

$843.20 is 7.75% of $10,880.

Exercise Set 7.4

1. What is 82% of 74?

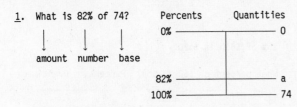

$$\frac{82}{100} = \frac{a}{74}$$

3. 4.3 is what percent of 5.9?

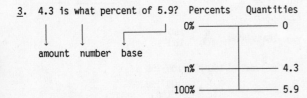

$$\frac{n}{100} = \frac{4.3}{5.9}$$

5. 14 is 25% of what?

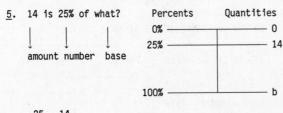

$$\frac{25}{100} = \frac{14}{b}$$

7. What is 84% of $50?

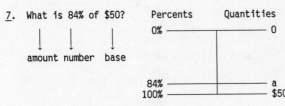

Translate: $\dfrac{84}{100} = \dfrac{a}{50}$

7. (continued)

Solve: 84·50 = 100·a   Finding cross products

$$\frac{84 \cdot 50}{100} = a$$ Dividing by 100

$$\frac{4200}{100} = a$$

$$42 = a$$

$42 is 84% of $50.

9. 80% of 550 is what?

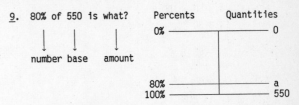

Translate: $\dfrac{80}{100} = \dfrac{a}{550}$

80·550 = 100·a   Finding cross products

$$\frac{80 \cdot 550}{100} = a$$ Dividing by 100

$$\frac{44,000}{100} = a$$

$$440 = a$$

80% of 550 is 440.

11. What is 8% of 1000?

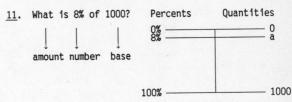

Translate: $\dfrac{8}{100} = \dfrac{a}{1000}$

Solve: 8·1000 = 100·a

$$\frac{8 \cdot 1000}{100} = a$$

$$\frac{8000}{100} = a$$

$$80 = a$$

80 is 8% of 1000.

13. 4.8% of 60 is what?   Percents   Quantities

$\downarrow$ $\downarrow$ $\downarrow$

number   base   amount

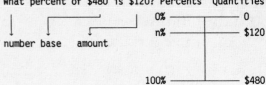

Translate: $\dfrac{4.8}{100} = \dfrac{a}{60}$

Solve: $4.8 \cdot 60 = 100 \cdot a$

$\dfrac{4.8 \cdot 60}{100} = a$

$\dfrac{288}{100} = a$

$2.88 = a$

4.8% of 60 is 2.88.

15. $24 is what percent of $96?   Percents   Quantities

$\downarrow$ $\downarrow$ $\downarrow$

amount   number   base

Translate: $\dfrac{n}{100} = \dfrac{24}{96}$

Solve: $96 \cdot n = 100 \cdot 24$

$n = \dfrac{100 \cdot 24}{96}$

$n = \dfrac{2400}{96}$

$n = 25$

$24 is 25% of $96.

17. 102 is what percent of 100?   Percents   Quantities

$\downarrow$ $\downarrow$ $\downarrow$

amount   number   base

Translate: $\dfrac{n}{100} = \dfrac{102}{100}$

Solve: $100 \cdot n = 100 \cdot 102$

$n = \dfrac{100 \cdot 102}{100}$

$n = \dfrac{10,200}{100}$

$n = 102$

102 is 102% of 100.

19. What percent of $480 is $120?   Percents   Quantities

$\downarrow$ $\downarrow$ $\downarrow$

number   base   amount

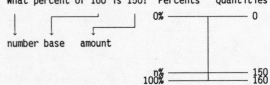

Translate: $\dfrac{n}{100} = \dfrac{120}{480}$

Solve: $480 \cdot n = 100 \cdot 120$

$n = \dfrac{100 \cdot 120}{480}$

$n = \dfrac{12,000}{480}$

$n = 25$

25% of $480 is $120.

21. What percent of 160 is 150?   Percents   Quantities

$\downarrow$ $\downarrow$ $\downarrow$

number   base   amount

Translate: $\dfrac{n}{100} = \dfrac{150}{160}$

Solve: $160 \cdot n = 100 \cdot 150$

$n = \dfrac{100 \cdot 150}{160}$

$n = \dfrac{15,000}{160}$

$n = 93.75$

93.75% $\left(\text{or } 93\tfrac{3}{4}\%\right)$ of 160 is 150.

23. $18 is 25% of what?   Percents   Quantities

$\downarrow$ $\downarrow$ $\downarrow$

amount   number   base

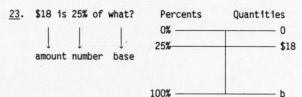

Translate: $\dfrac{25}{100} = \dfrac{18}{b}$

Solve: $25 \cdot b = 100 \cdot 18$

$b = \dfrac{100 \cdot 18}{25}$

$b = \dfrac{1800}{25}$

$b = 72$

$18 is 25% of $72.

25. 60% of what is 54?

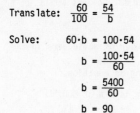

Percents   Quantities

$$0\% \rule{2cm}{0.4pt} 0$$
$$60\% \rule{2cm}{0.4pt} 54$$
$$100\% \rule{2cm}{0.4pt} b$$

Translate: $\dfrac{60}{100} = \dfrac{54}{b}$

Solve: $60 \cdot b = 100 \cdot 54$

$$b = \frac{100 \cdot 54}{60}$$

$$b = \frac{5400}{60}$$

$$b = 90$$

60% of 90 is 54.

27. 65.12 is 74% of what?

Percents   Quantities

$$0\% \rule{2cm}{0.4pt} 0$$
$$74\% \rule{2cm}{0.4pt} 65.12$$
$$100\% \rule{2cm}{0.4pt} b$$

amount number base

Translate: $\dfrac{74}{100} = \dfrac{65.12}{b}$

Solve: $74 \cdot b = 100 \cdot 65.12$

$$b = \frac{100 \cdot 65.12}{74}$$

$$b = \frac{6512}{74}$$

$$b = 88$$

65.12 is 74% of 88.

29. 80% of what is 16?

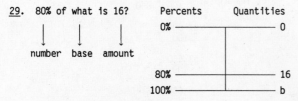

Percents   Quantities

$$0\% \rule{2cm}{0.4pt} 0$$
$$80\% \rule{2cm}{0.4pt} 16$$
$$100\% \rule{2cm}{0.4pt} b$$

Translate: $\dfrac{80}{100} = \dfrac{16}{b}$

Solve: $80 \cdot b = 100 \cdot 16$

$$b = \frac{100 \cdot 16}{80}$$

$$b = \frac{1600}{80}$$

$$b = 20$$

80% of 20 is 16.

31. What is $62\frac{1}{2}\%$ of 40?

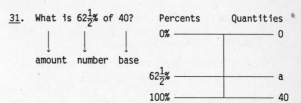

Percents   Quantities

$$0\% \rule{2cm}{0.4pt} 0$$
$$62\frac{1}{2}\% \rule{2cm}{0.4pt} a$$
$$100\% \rule{2cm}{0.4pt} 40$$

Translate: $\dfrac{62\frac{1}{2}}{100} = \dfrac{a}{40}$

Solve: $62\frac{1}{2} \cdot 40 = 100 \cdot a$

$$\frac{125}{2} \cdot \frac{40}{1} = 100 \cdot a$$

$$2500 = 100 \cdot a$$

$$\frac{2500}{100} = a$$

$$25 = a$$

25 is $62\frac{1}{2}\%$ of 40.

33. What is 9.4% of $8300?

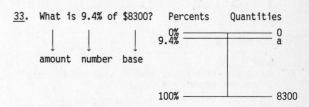

Percents   Quantities

$$0\% \rule{2cm}{0.4pt} 0$$
$$9.4\% \rule{2cm}{0.4pt} a$$
$$100\% \rule{2cm}{0.4pt} 8300$$

amount number base

Translate: $\dfrac{9.4}{100} = \dfrac{a}{8300}$

Solve: $9.4 \cdot 8300 = 100 \cdot a$

$$78{,}020 = 100 \cdot a$$

$$\frac{78{,}020}{100} = a$$

$$780.2 = a$$

$780.20 is 9.4% of $8300.

35. Estimate: Round 8.85% to 9% and $12,640 to $12,600. Then translate:

What is 9% of $12,600?

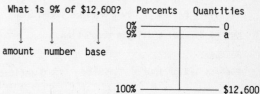

Percents   Quantities

$$0\% \rule{2cm}{0.4pt} 0$$
$$9\% \rule{2cm}{0.4pt} a$$
$$100\% \rule{2cm}{0.4pt} \$12{,}600$$

amount number base

$$\frac{9}{100} = \frac{a}{12{,}600}$$

Solve: $9 \cdot 12{,}600 = 100 \cdot a$

$$\frac{9 \cdot 12{,}600}{100} = a$$

$$\frac{113{,}400}{100} = a$$

$$1134 = a$$

$1134 is about 8.85% of $12,640. (Answers may vary.)

35. (continued)

Calculate: First we translate.

What is 8.85% of $12,640? Percents   Quantities

  ↓   ↓   ↓   0% ————————— 0

              8.85% ————————— a

amount number base

100% ————————— $12,640

$$\frac{8.85}{100} = \frac{a}{12,640}$$

Solve: $8.85 \cdot 12{,}640 = 100 \cdot a$

$$\frac{8.85 \cdot 12{,}640}{100} = a$$

$$\frac{111{,}864}{100} = a \qquad \text{Use a calculator to multiply and divide.}$$

$$1118.64 = a$$

$1118.64 is 8.85% of $12,640.

Exercise Set 7.5

1. We first find how many bowlers would be expected to be left-handed.

Method 1: Solve using an equation.

    Restate: What is 17% of 160?

                  ↓   ↓ ↓ ↓ ↓

    Translate: a = 17% × 160

This tells us what to do. We convert 17% to decimal notation and multiply.

```
   1 6 0
 × 0.1 7      (17% = 0.17)
   1 1 2 0
   1 6 0 0
   2 7.2 0
```

You would expect 27 bowlers to be left-handed.

Method 2: Solve using a proportion.

Restate:

What is 17% of 160?   Percents   Quantities

 ↓   ↓   ↓   0% ————————— 0

amount number base   17% ————————— a

100% ————————— 160

    Translate: $\dfrac{17}{100} = \dfrac{a}{160}$

    Solve: $17 \cdot 160 = 100 \cdot a$

            $\dfrac{17 \cdot 160}{100} = a$

            $\dfrac{2720}{100} = a$

            $27.2 = a$

1. (continued)

You would expect 27 bowlers to be left-handed.

We can subtract to find the number of bowlers that would not be left-handed.

    $160 - 27 = 133$

You would expect that 133 bowlers would not be left-handed.

3. We first find the number of moviegoers in the 12-29 age group.

Method 1: Solve using an equation.

    Restate: What is 67% of 800?

               ↓ ↓ ↓ ↓ ↓

    Translate: a = 67% × 800

This tells us what to do. We convert 67% to decimal notation and multiply.

```
     8 0 0
   × 0.6 7      (67% = 0.67)
     5 6 0 0
   4 8 0 0 0
   5 3 6.0 0
```

Thus, 536 were in the 12-29 age group.

Method 2: Solve using a proportion.

Restate:

What is 67% of 800?   Percents   Quantities

 ↓   ↓   ↓   0% ————————— 0

amount number base   67% ————————— a

100% ————————— 800

    Translate: $\dfrac{67}{100} = \dfrac{a}{800}$

            $\dfrac{67 \cdot 800}{100} = a$

            $\dfrac{53{,}600}{100} = a$

            $536 = a$

Thus, 536 were in the 12-29 age group.

We subtract to find the number that were not in this age group.

    $800 - 536 = 264$

264 moviegoers were not in the 12-29 age group.

5. We first find the percent that are hits.

Method 1: Solve using an equation.

Restate: 13 is what percent of 40?

Translate: 13 =   n   × 40

To solve the equation we divide on both sides by 40:

$$13 \div 40 = n$$
$$0.325 = n$$
$$32.5\% = n$$

Thus, 32.5% are hits.

Method 2: Solve using a proportion.

Restate:

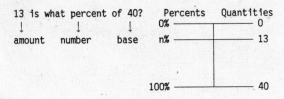

13 is what percent of 40?
amount   number   base

Percents   Quantities
0% ———————— 0
n% ———————— 13

100% ———————— 40

Translate: $\dfrac{n}{100} = \dfrac{13}{40}$

Solve: $40 \cdot n = 100 \cdot 13$

$$n = \frac{100 \cdot 13}{40}$$
$$n = \frac{1300}{40}$$
$$n = 32.5$$

Thus, 32.5% are hits.

We subtract to find the percent that are not hits.

$$100\% - 32.5\% = 67.5\%$$

Thus, 67.5% are not hits.

7. We first find how many milliliters are acid.

Method 1: Solve using an equation.

Restate: What is 3% of 680?

Translate: a = 3% × 680

This tells us what to do. We convert 3% to decimal notation and multiply.

```
  6 8 0
× 0.0 3      (3% = 0.03)
2 0.4 0
```

Thus, 20.4 milliliters are acid.

7. (continued)

Method 2: Solve using a proportion.

Restate:

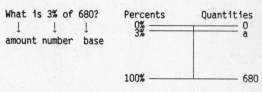

What is 3% of 680?
amount number base

Percents   Quantities
0% ———————— 0
3% ———————— a

100% ———————— 680

Translate: $\dfrac{3}{100} = \dfrac{a}{680}$

Solve: $680 \cdot 3 = 100 \cdot a$

$$\frac{680 \cdot 3}{100} = a$$
$$\frac{2040}{100} = a$$
$$20.4 = a$$

Thus, 20.4 milliliters are acid.

We subtract to find how many milliliters are water.

$$680 - 20.4 = 659.6$$

Thus, 659.6 milliliters are water.

9. Method 1: Solve using an equation.

Restate: 2190 is what percent of 8760?

Translate: 2190 = n   %   × 8760

$$2190 \div 8760 = n$$
$$0.25 = n$$
$$25\% = n$$

Thus, 2190 is 25% of 8760.

Method 2: Solve using a proportion.

Restate:

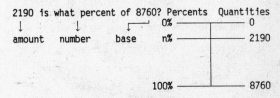

2190 is what percent of 8760?
amount   number   base

Percents   Quantities
0% ———————— 0
n% ———————— 2190

100% ———————— 8760

Translate: $\dfrac{n}{100} = \dfrac{2190}{8760}$

Solve: $8760 \cdot n = 100 \cdot 2190$

$$n = \frac{100 \cdot 2190}{8760}$$
$$n = \frac{219,000}{8760}$$
$$n = 25$$

Thus, 2190 is 25% of 8760.

11. First we find the total weight of the nuts.

    1800 lb + 1500 lb + 700 lb = 4000 lb

Then we find what percent are peanuts.

<u>Method 1</u>: Solve using an equation.

Restate: <u>What percent</u> of 4000 is 1800?

Translate: n   × 4000 = 1800

    n = 1800 ÷ 4000

    n = 0.45

    n = 45%

Thus, 45% are peanuts.

<u>Method 2</u>: Solve using a proportion.

Restate: What percent of 4000 is 1800?

          number   base   amount

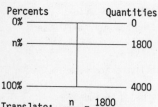

Translate: $\dfrac{n}{100} = \dfrac{1800}{4000}$

Solve: $4000 \cdot n = 100 \cdot 1800$

$$n = \frac{100 \cdot 1800}{4000}$$

$$n = \frac{180{,}000}{4000}$$

$$n = 45$$

Thus, 45% are peanuts.

Next we find the percent that are cashews.

<u>Method 1</u>: Solve using an equation.

Restate: <u>What percent</u> of 4000 is 1500?

Translate: n   × 4000 = 1500

    n = 1500 ÷ 4000

    n = 0.375

    n = 37.5%

Thus, 37.5% are cashews.

<u>Method 2</u>: Solve using a proportion.

Restate: What percent of 4000 is 1500?

          number   base   amount

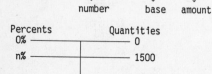

---

11. (continued)

Translate: $\dfrac{n}{100} = \dfrac{1500}{4000}$

Solve: $4000 \cdot n = 100 \cdot 1500$

$$n = \frac{100 \cdot 1500}{4000}$$

$$n = \frac{150{,}000}{4000}$$

$$n = 37.5$$

Thus, 37.5% are cashews.

Finally, we find the percent that are almonds. First we observe that peanuts and cashews make up 45% + 37.5%, or 82.5%, of the total weight. Then we subtract.

    100% - 82.5% = 17.5%

Thus, 17.5% are almonds.

13.

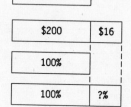

First find the increase by subtracting.

    2 1 6   New amount

   - 2 0 0   Original amount

     1 6   Increase

The increase is $16. Now we ask:

$16 is what percent of $200 (the original amount)?

<u>Method 1</u>: Solve using an equation.

$16 is <u>what percent</u> of $200?

16 =   n   × 200   Translating

16 ÷ 200 = n   Dividing on both sides by 200

  0.08 = n

    8% = n

The percent of increase was 8%.

13. (continued)

Method 2: Solve using a proportion.

$16 is what percent of $200?
  ↓   ↓   ↓
amount   number   base

$$\frac{n}{100} = \frac{16}{200} \qquad \text{Translating}$$

$$200 \cdot n = 16 \cdot 100$$

$$n = \frac{16 \cdot 100}{200}$$

$$n = \frac{1600}{200}$$

$$n = 8$$

The percent of increase was 8%.

15.

$70

$56	$14

100%

	?%

First find the decrease by subtracting.

```
  7 0
- 5 6
  1 4
```

The decrease is $14. Now we ask:

$14 is what percent of $70 (the original price)?

Method 1: Solve using an equation.

$14 is what percent of $70?
 ↓   ↓   ↓   ↓   ↓
14   =   n   ×   70

$$14 \div 70 = n \qquad \text{Dividing on both sides by 70}$$

$$0.2 = n$$

$$20\% = n$$

The percent of decrease was 20%.

15. (continued)

Method 2: Solve using a proportion.

$14 is what percent of $70?
  ↓   ↓   ↓
amount   number   base

$$\frac{n}{100} = \frac{14}{70}$$

$$70 \cdot n = 100 \cdot 14 \qquad \text{Finding cross products}$$

$$n = \frac{100 \cdot 14}{70} \qquad \text{Dividing by 70}$$

$$n = \frac{1400}{70}$$

$$n = 20$$

The percent of decrease was 20%.

17.

$8600	$?

100%	5%

First, find the increase. We ask:

What is 5% of $8600?

Method 1: Solve using an equation.

What is 5% of $8600?
 ↓   ↓ ↓ ↓   ↓
 a   = 5% × 8600

We convert to decimal notation and multiply:

```
  8 6 0 0
× 0.0 5
4 3 0.0 0
```

The increase is $430.

Method 2: Solve using a proportion.

What is 5% of $8600?
  ↓   ↓   ↓
amount   number   base

$$\frac{5}{100} = \frac{a}{8600}$$

$$5 \cdot 8600 = 100 \cdot a$$

$$\frac{5 \cdot 8600}{100} = a$$

$$\frac{43,000}{100} = a$$

$$430 = a$$

The increase is $430.

The new salary is $8600 + 430 = $9030.

19.

| $12,000 |

| | $? |

| 100% |

| 70% | 30% |

First, find the decrease. We ask:

What is 30% of $12,000?

Method 1: Solve using an equation.

What is 30% of $12,000?
↓ ↓ ↓ ↓ ↓
a = 30% × $12,000

Convert 30% to decimal notation and multiply.

```
  1 2,0 0 0
×     0.3
  3 6 0 0.0
```

The decrease is $3600.

Method 2: Solve using a proportion.

What is 30% of $12,000?
↓ ↓ ↓
amount number base

$$\frac{30}{100} = \frac{a}{12,000}$$

$$30 \cdot 12,000 = 100 \cdot a$$

$$\frac{30 \cdot 12,000}{100} = a$$

$$\frac{360,000}{100} = a$$

$$3600 = a$$

The decrease is $3600.

The value 1 year later is $12,000 − $3600 = $8400.

21. To find the population in 1991 we first find the increase. We ask:

What is 1.6% of 5.2 billion?

Method 1: Solve using an equation.

What is 1.6% of 5.2 billion?
↓ ↓ ↓ ↓ ↓
a = 1.6% × 5.2

We convert 1.6% to decimal notation and multiply.

```
    5.2
× 0.0 1 6
    3 1 2
    5 2 0
0.0 8 3 2
```

21. (continued)

The increase is 0.0832 billion.

Method 2: Solve using a proportion.

What is 1.6% of 5.2 billion?
↓ ↓ ↓
amount number base

$$\frac{1.6}{100} = \frac{a}{5.2}$$

$$1.6 × 5.2 = 100 × a$$

$$\frac{1.6 × 5.2}{100} = a$$

$$\frac{8.32}{100} = a$$

$$0.0832 = a$$

The increase is 0.0832 billion.

To find the population in 1991 we add:
5.2 + 0.0832 = 5.2832
The population in 1991 will be 5.2832 billion.

To find the population in 1992 we first find the increase. We ask:

What is 1.6% of 5.2832 billion?

Method 1: Solve using an equation.

What is 1.6% of 5.2832 billion?
↓ ↓ ↓ ↓ ↓
a = 1.6% × 5.2832

We convert 1.6% to decimal notation and multiply.

```
    5.2 8 3 2
×     0.0 1 6
    3 1 6 9 9 2
    5 2 8 3 2 0
0.0 8 4 5 3 1 2
```

The increase is 0.0845312 billion.

Method 2: Solve using a proportion.

What is 1.6% of 5.2832 billion?
↓ ↓ ↓
amount number base

$$\frac{1.6}{100} = \frac{a}{5.2832}$$

$$1.6 × 5.2832 = 100 × a$$

$$\frac{1.6 × 5.2832}{100} = a$$

$$\frac{8.45312}{100} = a$$

$$0.0845312 = a$$

The increase is 0.0845312 billion.

21. (continued)

To find the population in 1992 we add:

5.2832 + 0.0845312 = 5.3677312

The population in 1992 will be about 5.3677 billion.

To find the population in 1993, we first find the increase. We ask:

What is 1.6% of 5.3677 billion?

Method 1: Solve using an equation.

What is 1.6% of 5.3677 billion?
↓   ↓ ↓   ↓   ↓
a   = 1.6% ×   5.3677

We convert 1.6% to decimal notation and multiply.

$$\begin{array}{r} 5.3677 \\ \times\ \ 0.016 \\ \hline 322062 \\ 53677\ 0 \\ \hline 0.0858832 \end{array}$$

The increase is 0.0858832.

Method 2: Solve using a proportion.

What is 1.6% of 5.3677 billion?
↓   ↓   ↓
amount number   base

$$\frac{1.6}{100} = \frac{a}{5.3677}$$

$$1.6 \times 5.3677 = 100 \times a$$

$$\frac{1.6 \times 5.3677}{100} = a$$

$$\frac{8.58832}{100} = a$$

$$0.0858832 = a$$

The increase is 0.0858832.

To find the population in 1993 we add:

5.3677 + 0.0858832 = 5.4535832

The population in 1993 will be about 5.4536 billion.

23. Since the car depreciates 30% in the first year, its value after the first year is 100% - 30%, or 70%, of the original value. So we ask:

$8750 is 70% of what?

Method 1: Solve using an equation.

8750 is 70% of what?
↓   ↓ ↓   ↓ ↓
8750 = 70% ×   b

$$8750 = 0.7 \times b$$

$$8750 \div 0.7 = b$$

$$12,500 = b$$

The original cost was $12,500.

23. (continued)

Method 2: Solve using a proportion.

8750 is 70% of what?
↓   ↓   ↓
amount number base

$$\frac{70}{100} = \frac{8750}{b}$$

$$70 \cdot b = 100 \cdot 8750$$

$$b = \frac{100 \cdot 8750}{70}$$

$$b = \frac{875,000}{70}$$

$$b = 12,500$$

The original cost was $12,500.

25. At age 25: First we subtract.

220 - 25 = 195

Then we multiply by 85%. We convert 85% to decimal notation.

195 × 0.85 = 165.75 ≈ 166   (Rounding)

The maximal heartbeat of a person of age 25 is 166 beats per minute.

At age 36: First we subtract.

220 - 36 = 184

Then we multiply.

184 × 0.85 = 156.4 ≈ 156

The maximal heartbeat of a person of age 36 is 156 beats per minute.

At age 48: First we subtract.

220 - 48 = 172

Then we multiply.

172 × 0.85 = 146.2 ≈ 146

The maximal heartbeat for a person of age 48 is 146 beats per minute.

At age 60: First we subtract.

220 - 60 = 160

Then we multiply.

160 × 0.85 = 136

The maximal heartbeat for a person of age 60 is 136 beats per minute.

At age 76: First we subtract.

220 - 76 = 144

Then we multiply.

144 × 0.85 = 122.4 ≈ 122

The maximal heartbeat for a person of age 76 is 122 beats per minute.

27. a) We find $1000 increased by 15%. First we find
the amount of the increase.

What is 15% of $1000?
↓   ↓   ↓   ↓   ↓
a   = 15% ×   1000

We multiply.

0.15 × 1000 = 150

The increase is $150. Then $1000 increased by
15% is $1000 + $150, or $1150.

Next we find $1150 decreased by 15%. First we
find the amount of the decrease.

What is 15% of $1150?
↓   ↓   ↓   ↓   ↓
a   = 15% ×   1150

We multiply.

0.15 × 1150 = 172.5

The decrease is $172.50. Then $1150 decreased
by 15% is $1150 - $172.50, or $977.50.

b) We find $1000 decreased by 15%. First we
find the amount of the decrease, or 15% of
$1000. In part (a) we found this to be $150.
Then $1000 decreased by 15% is $1000 - $150,
or $850.

Next we find $850 increased by 15%. First we
find the amount of the increase.

What is 15% of $850?
↓   ↓   ↓   ↓   ↓
a   = 15% ×   850

We multiply.

0.15 × 850 = 127.5

The increase is $127.50. Then $850 increased
by 15% is $850 + $127.50, or $977.50.

Neither (a) nor (b) is higher. They are the same.

29. Note that 4 ft, 8 in. = 48 in. + 8 in. = 56 in.

56 inches is 84.4% of what?
↓   ↓   ↓   ↓   ↓
56   = 84.4% ×   b

To solve the equation we divide on both sides by
84.4%:

56 ÷ 84.4% = b

56 ÷ 0.844 = b

66 ≈ b

```
              6 6. 3
0.8 4 4∧⟌5 6.0 0 0∧0
        5 0 6 4
          5 3 6 0
          5 0 6 4
            2 9 6 0
            2 5 3 2
              4 2 8
```

Her final adult height is about 66 in. or 5 ft,
6 in.

1. a) We first find the sales tax. It is

8.25% of $248, or 0.0825 × 248,
which is $20.46.

b) The total price is purchase
price plus sales tax, or

2 4 8.0 0
+ 2 0.4 6
2 6 8.4 6

```
        2 4 8
    × 0.0 8 2 5
      1 2 4 0
      4 9 6 0
    1 9 8 4 0 0
    2 0.4 6 0 0
```

The total price is $268.46.

3. a) We first find the sales tax. It is

6% of $189.95, or 0.06 × 189.95,
which is 11.397 or about $11.40.

```
    1 8 9.9 5
    × 0.0 6
  1 1.3 9 7 0
```

b) The total price is purchase price plus sales
tax, or

1 8 9.9 5
+ 1 1.4 0
2 0 1.3 5

The total price is $201.35.

5. Think:

Sales tax	is	what percent	of	purchase price?
↓	↓	↓	↓	↓

Translate: 48   =   r   ×   960

To solve the equation we divide on both sides by
960:

48 ÷ 960 = r

0.05 = r

5% = r

The sales tax rate is 5%.

7. Think:

Sales tax	is	what percent	of	purchase price?
↓	↓	↓	↓	↓

Translate: 35.80   =   r   ×   895

To solve the equation we divide on both sides by
895:

35.80 ÷ 895 = r

0.04 = r

4% = r

The sales tax rate is 4%.

9. Think: Sales tax is 6% of what?

Translate: $168 = 6\% \times p$, or $168 = 0.06 \times p$

To solve the equation we divide on both sides by 0.06:

$168 \div 0.06 = p$

$2800 = p$

```
              2 8 0 0.
0.0 6∧⟌1 6 8.0 0∧
       1 2 0 0 0
         4 8 0 0
         4 8 0 0
               0
               0
               0
```

The purchase price is $2800.

11. Think: Sales tax is 3.5% of what?

Translate: $28 = 3.5\% \times p$, or $28 = 0.035 \times p$

To solve the equation we divide on both sides by 0.035:

$28 \div 0.035 = p$

$800 = p$

```
              8 0 0.
0.0 3 5∧⟌2 8.0 0 0∧
         2 8 0 0 0
                 0
                 0
                 0
```

The purchase price is $800.

13. The total tax rate is city tax rate plus state tax rate, or 1% + 4% = 5%.

Find the sales tax. It is

5% of $665, or 0.05×665,

which is $33.25.

```
    6 6 5
  × 0.0 5
  3 3.2 5
```

The tax is $33.25.

15. Think:

Sales tax is what percent of purchase price?

Translate: $1030.40 = r \times 18{,}400$

To solve the equation we divide on both sides by 18,400:

$1030.40 \div 18{,}400 = r$

$0.056 = r$

$5.6\% = r$

The sales tax rate is 5.6%.

17. $2.3 \times y = 85.1$

$y = 85.1 \div 2.3$

$y = 37$

```
            3 7.
2.3∧⟌8 5.1∧
     6 9 0
     1 6 1
     1 6 1
         0
```

19. $\dfrac{13}{11} = 13 \div 11$

```
       1.1 8 1 8        We get a repeating decimal.
11⟌1 3.0 0 0 0
   1 1
     2 0
     1 1
       9 0
       8 8
         2 0
         1 1
           9 0
           8 8
             2
```

$\dfrac{13}{11} = 1.\overline{18}$

21. The sales tax is

5.4% of $96,568.95 or $0.054 \times 96{,}568.95$.

Use a calculator to carry out the multiplication.

$0.054 \times 96{,}568.95 = 5214.7233$ or about 5214.72

The sales tax is $5214.72.

Exercise Set 7.7

1. Commission = Commission rate × Sales

   C   =   20%   × 18,450

   This tells us what to do. We multiply.

```
  1 8,4 5 0
  ×    0.2      (20% = 0.20 = 0.2)
  3 6 9 0.0
```

   The commission is $3690.

3. Commission = Commission rate × Sales

   120   =   r   × 2400

   To solve this equation we divide on both sides by 2400:

   $120 \div 2400 = r$

   We can divide, but this time we simplify by removing a factor of 1:

   $r = \dfrac{120}{2400} = \dfrac{1}{20} \cdot \dfrac{120}{120} = \dfrac{1}{20} = 0.05 = 5\%$

   The commission rate is 5%.

5. Commission = Commision rate × Sales
   392   =   40%   ×   S

To solve this equation we divide on both sides by 40%:

   392 ÷ 40% = S

   392 ÷ 0.4 = S   (40% = 0.40 = 0.4)

      980 = S

$$0.4_\wedge \overline{)3\,9\,2.0_\wedge}$$

```
           9 8 0.
0.4∧ )3 9 2.0∧
      3 6 0 0
        3 2 0
        3 2 0
            0
            0
            0
```

There were $980 worth of sweepers sold.

7. Commission = Commission rate × Sales
   C   =   7%   × 98,000

This tells us what to do. We multiply:

```
   9 8,0 0 0
   ×  0.0 7
   6 8 6 0.0 0
```

The commission is $6860.

9. First we find the commission.

Commission = Commission rate × Sales
   C   =   2%   × 990

This tells us what to do. We multiply:

```
   9 9 0
   × 0.0 2     (2% = 0.02)
   1 9.8 0
```

The commission is $19.80.

The total wages were salary plus commission, or

```
   5 0 0.0 0
   +  1 9.8 0
   5 1 9.8 0
```

The total wages for that month were $519.80.

11. Discount = Rate of discount × Marked price
       ↓   ↓   ↓   ↓   ↓
       D   =   10%   × $300

Convert 10% to decimal notation and multiply.

```
    3 0 0
   ×  0.1     (10% = 0.10 = 0.1)
    3 0.0
```

The discount is $30.

11. (continued)

   Sale price = Marked price - Discount
      S   =   300   -   30

   We subtract: 3 0 0
                  - 3 0
                  2 7 0

The sale price is $270.

13. Discount = Rate of discount × Marked price
       D   =   60%   ×   5

Convert 60% to decimal notation and multiply.

```
      5
   × 0.6      (60% = 0.60 = 0.6)
    3.0
```

The discount is $3.

Sale price = Marked price - Discount
   S   =   5   -   3

We subtract: 5 - 3 = 2

The sale price is $2.

15. Discount = Rate of discount × Marked price
       D   =   10%   ×   125

Convert 10% to decimal notation and multiply.

```
    1 2 5
   ×  0.1
    1 2.5
```

The discount is $12.50.

Sale price = Marked priced - Discount
   S   =   125.00   -   12.50

We subtract:

```
    1 2 5.0 0
   -  1 2.5 0
    1 1 2.5 0
```

The sale price is $112.50.

17. Discount = Rate of discount × Marked price

     240   =   R   ×   600

To solve the equation we divide on both sides by R:

     240 ÷ 600 = R

We can simplify by removing a factor of 1:

$R = \frac{240}{600} = \frac{2}{5} \cdot \frac{120}{120} = \frac{2}{5} = 0.4 = 40\%$

The rate of discount is 40%.

Sale price = Marked price - Discount

     S   =   600   -   240

We subtract:

```
   6 0 0
 - 2 4 0
   3 6 0
```

The sale price is $360.

19. Discount = Marked price - Sale price

     D   =   1275   -   888

We subtract:

```
   1 2 7 5
 -   8 8 8
     3 8 7
```

The discount is $387.

Discount = Rate of discount × Marked price

     387   =   R   ×   1275

To solve the equation we divide on both sides by 1275:

     387 ÷ 1275 = R
     0.30353 ≈ R
     30.353% ≈ R, or
     $30\frac{6}{17}\%$ = R

The commission rate is $30\frac{6}{17}\%$, or about 30.353%.

21. $\frac{5}{9}$ = 5 ÷ 9

```
    0.5 5      We get a repeating decimal.
 9)5.0 0
   4 5
     5 0
     4 5
       5
```

$\frac{5}{9} = 0.\overline{5}$

23. $\frac{11}{12}$ = 11 ÷ 12

```
      0.9 1 6 6     We get a repeating decimal.
 1 2)1 1.0 0 0 0
     1 0 8
       2 0
       1 2
         8 0
         7 2
           8 0
           7 2
             8
```

$\frac{11}{12} = 0.91\overline{6}$

25. Commission = Commission rate × Sales

     C   =   7.5%   × 78,990

We multiply.

```
     7 8,9 9 0
   ×   0.0 7 5      (7.5% = 0.075)
     3 9 4 9 5 0
   5 5 2 9 3 0 0
   5 9 2 4.2 5 0
```

The commission is $5924.25.

We subtract to find how much the seller gets for the house after paying the commission.

     $78,990 - $5924.25 = $73,065.75

Exercise Set 7.8

1. We take 13% of $200:

     13% × 200 = 0.13 × 200   2 0 0
              = 26   × 0.1 3
                                     6 0 0
                                   2 0 0 0
                                   2 6.0 0

The interest is $26.

3. We take 12.4% of $2000:

     12.4% × 2000 = 0.124 × 2000   2 0 0 0
                 = 248   × 0.1 2 4
                                        8 0 0 0
                                      4 0 0 0 0
                                    2 0 0 0 0 0
                                    2 4 8.0 0 0

The interest is $248.

5. Interest = (Interest for 1 year) $\times \frac{1}{4}$

    $= (14\% \times \$4300) \times \frac{1}{4}$

    $= 0.14 \times 4300 \times \frac{1}{4}$

    $= 602 \times \frac{1}{4}$

```
    4 3 0 0
  ×   0.1 4
    1 7 2 0 0
    4 3 0 0 0
    6 0 2.0 0
```

    $= \frac{602}{4}$

```
        1 5 0.5
    4 ⟌6 0 2.0
      4 0 0
        2 0 2
        2 0 0
          2 0
          2 0
            0
```

    $= 150.5$

The interest is $150.50.

7. Interest = (Interest for 1 year) $\times \frac{60}{360}$

    $= (14.5\% \times \$5000) \times \frac{60}{360}$

    $= 0.145 \times 5000 \times \frac{1}{6}$ $\frac{60}{360} = \frac{1}{6}$

    $= 725 \times \frac{1}{6}$

    $= \frac{725}{6}$

```
        1 2 0.8 3 3
    6 ⟌7 2 5.0 0 0
      6 0 0
      1 2 5
      1 2 0
        5 0
        4 8
        2 0
        1 8
        2 0
        1 8
          2
```

    $= 120.83\overline{3}$

    $= 120.83$ (Rounded to the nearest hundredth)

The interest is $120.83.

9. We take 9.4% of $4400:

    9.4% of $4400 $= 0.094 \times 4400$

    $= 413.6$

```
      4 4 0 0
    ×   0.0 9 4
      1 7 6 0 0
      3 9 6 0 0 0
      4 1 3.6 0 0
```

The interest is $413.60.

11. a) Find the interest at the end of 1 year.

        I $= 10\% \times \$400$

        $= 0.1 \times \$400$

        $= \$40$

    b) Find the principal after 1 year.

        $\$400 + \$40 = \$440$

    c) Going into the second year the principal is $440. Find the interest for 1 year after that.

        I $= 10\% \times \$440$

        $= 0.1 \times \$440$

        $= \$44$

    d) Find the new principal after 2 years.

        $\$440 + \$44 = \$484$

    The amount in the account after 2 years is $484.

13. a) Find the interest at the end of 1 year.

        I $= 8.8\% \times \$200$

        $= 0.088 \times \$200$

        $= 17.60$

```
          2 0 0
        × 0.0 8 8
          1 6 0 0
        1 6 0 0 0
        1 7.6 0 0
```

    b) Find the principal after 1 year.

        $\$200.00 + \$17.60 = \$217.60$

    c) Going into the second year the principal is $217.60. Find the interest for 1 year after that.

        I $= 8.8\% \times \$217.60$

        $= 0.088 \times \$217.60$

        $= \$19.1488$

```
          2 1 7.6
        ×   0.0 8 8
          1 7 4 0 8
        1 7 4 0 8 0
        1 9.1 4 8 8
```

        $= \$19.15$ Rounded to the nearest hundredth

    d) Find the new principal after 2 years.

        $\$217.60 + \$19.15 = \$236.75$

15. a) Find the interest at the end of $\frac{1}{2}$ year.

$$I = 16\% \times \$400 \times \frac{1}{2}$$

$$= 0.16 \times \$400 \times \frac{1}{2}$$

$$= \$64 \times \frac{1}{2}$$

$$= \$32$$

b) Find the new principal after $\frac{1}{2}$ year.

$$\$400 + \$32 = \$432$$

c) Going into the last half of the year the principal is $432. Find the interest for $\frac{1}{2}$ year after that.

$$I = 16\% \times \$432 \times \frac{1}{2}$$

$$= 0.16 \times \$432 \times \frac{1}{2}$$

$$= \$69.12 \times \frac{1}{2}$$

$$= \$34.56$$

d) Find the new principal after 1 year.

$$\$432 + \$34.56 = \$466.56$$

The amount in the account after 1 year is $466.56.

17. a) Find the interest at the end of $\frac{1}{2}$ year.

$$I = 9\% \times \$2000 \times \frac{1}{2}$$

$$= 0.09 \times \$2000 \times \frac{1}{2}$$

$$= \$180 \times \frac{1}{2}$$

$$= \$90$$

b) Find the new principal after $\frac{1}{2}$ year.

$$\$2000 + \$90 = \$2090$$

c) Going into the last half of the year the principal is $2090. Find the interest for $\frac{1}{2}$ year after that.

$$I = 9\% \times \$2090 \times \frac{1}{2}$$

$$= 0.09 \times \$2090 \times \frac{1}{2}$$

$$= \$188.10 \times \frac{1}{2}$$

$$= \$94.05$$

d) Find the new principal after 1 year.

$$\$2090 + \$94.05 = \$2184.05$$

19. $\dfrac{9}{10} = \dfrac{x}{5}$

$9 \cdot 5 = 10 \cdot x$

$\dfrac{9 \cdot 5}{10} = x$

$\dfrac{45}{10} = x$

$\dfrac{9}{2} = x$, or

$4.5 = x$

21. $\dfrac{100}{3} = 33\frac{1}{3}$

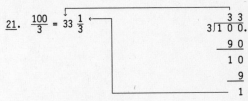

23. Interest = (Interest for 1 year) $\times \dfrac{3}{4}$

$$= (7.75\% \times \$24{,}680) \times \frac{3}{4}$$

$$= 0.0775 \times \$24{,}680 \times \frac{3}{4}$$

$$= \frac{0.0775 \times \$24{,}680 \times 3}{4}$$

$= \$1434.525$ Using a calculator to multiply and divide

$= \$1434.53$ Rounded to the nearest hundredth

The interest is $1434.53.

Exercise Set 8.1

1. To find the average, add the numbers. Then divide by the number of addends.

$$\frac{8 + 7 + 15 + 15 + 15 + 12}{6} = \frac{72}{6} = 12$$

The average is 12.

To find the median, first list the numbers in order from the smallest to largest.

7, 8, 12, 15, 15, 15

└── Median = 13.5

The median is halfway between 12 and 15.

We find it as follows:

$$\frac{12 + 15}{2} = \frac{27}{2} = 13.5$$

The median is 13.5.

Find the mode:

The number that occurs most often is 15. The mode is 15.

3. To find the average, add the numbers. Then divide by the number of addends.

$$\frac{5 + 10 + 15 + 20 + 25 + 30 + 35}{7} = \frac{140}{7} = 20$$

The average is 20.

Find the median:

5, 10, 15, 20 , 25, 30, 35

↑── Median is 20

The middle number is the median. Thus, 20 is the median of the set of numbers.

Find the mode:
No number repeats, so each number in the set is a mode. The modes are 5, 10, 15, 20, 25, 30, and 35.

5. Find the average:

$$\frac{1.2 + 4.3 + 5.7 + 7.4 + 7.4}{5} = \frac{26}{5} = 5.2$$

The average is 5.2.

Find the median:

1.2, 4.3, 5.7 , 7.4, 7.4

└── Median is 5.7

The middle number is the median. Thus, 5.7 is the median of the set of numbers.

Find the mode:
The number that occurs most often is 7.4 The mode is 7.4.

7. Find the average:

$$\frac{234 + 228 + 234 + 229 + 234 + 278}{6} = \frac{1437}{6} = 239.5$$

The average is 239.5.

Find the median:

228, 229, 234, 234, 234, 278

└── Median = 234

The median is halfway between 234 and 234. Although it seems clear that this is 234, we can compute it as follows:

$$\frac{234 + 234}{2} = \frac{468}{2} = 234$$

The median is 234.

Find the mode:
The number that occurs most often is 234. The mode is 234.

9. Find the average weight:

$$\frac{250 + 255 + 260 + 260}{4} = \frac{1025}{4} = 256.25$$

The average weight is 256.25 lb.

Find the median weight:

250, 255, 260, 260

└── Median is 257.5

The median is halfway between 255 and 260. We find it as follows:

$$\frac{255 + 260}{2} = \frac{515}{2} = 257.5$$

The median is 257.5 lb.

Find the mode:

The number that occurs most often is 260. The mode is 260 lb.

11. Find the average:

$$\frac{43° + 40° + 23° + 38° + 54° + 35° + 47°}{7} = \frac{280°}{7} = 40°$$

The average temperature was 40°.

Find the median temperature.

List the temperatures in order:

23°, 35°, 38°, 40° , 43°, 47°, 54°

↑── Median is 40°

The middle number is called the median.

Thus, 40° is the median temperature.

Find the mode:

No number repeats, so each number in the set is a mode. The modes are 43°, 40°, 23°, 38°, 54°, 35°, and 47°.

13. We divide the total number of miles, 522, by the number of gallons, 18.

$$\frac{522}{18} = 29$$

The average was 29 miles per gallon.

15. To find the GPA we first add the grade point values for each hour taken. This is done by first multiplying the grade point value by the number of hours in the course and then adding as follows:

B ——> 3.00·4 = 12
B ——> 3.00·5 = 15
B ——> 3.00·3 = 9
C ——> 2.00·4 = 8
        (Total) 44

The total number of hours taken is

4 + 5 + 3 + 4, or 16. We divide 44 by 16.

$$\frac{44}{16} = 2.75$$

The student's grade point average is 2.75.

17. Find the average price per pound:

$$\frac{\$9.79 + \$9.59 + \$9.69 + \$9.79 + 93.89}{5} = \frac{\$48.75}{5}$$

$$= \$9.75$$

The average price per pound of steak was $9.75.

Find the median price per pound:
List the prices in order:

$9.59, $9.69, 9.79 , $9.79, $9.89

                    Median is $9.79

The middle number is called the median.
Thus, $9.79 is the median price per pound of steak.

Find the mode:
The number that occurs most often is $9.79. The mode is $9.79.

19. We can find the total of the five scores needed as follows:

80 + 80 + 80 + 80 + 80 = 400.

The total of the scores on the first four tests is

80 + 74 + 81 + 75 = 310.

Thus the student needs to get at least

400 - 310, or 90

to get a B. We can check this as follows:

$$\frac{80 + 74 + 81 + 75 + 90}{5} = \frac{400}{5} = 80.$$

21. To find the average we first add all the salaries. We do this by multiplying each salary by the number of employees earning that salary as follows:

Owner ———> $29,200·1 = $ 2 9,2 0 0
Salesperson —> 19,600·5 = 9 8,0 0 0
Secretary ———> 14,800·3 = 4 4,4 0 0
Custodian ———> 13,000·1 = 1 3,0 0 0
                            $1 8 4,6 0 0 (Total)

The total number of employees is 1 + 5 + 3 + 1, or 10. We divide $184,600 by 10:

$$\frac{\$184,600}{10} = \$18,460.$$

The average salary is $18,460.

23. $\frac{2}{3} \cdot \frac{2}{3} = \frac{2 \cdot 2}{3 \cdot 3} = \frac{4}{9}$

25.
```
      1.4 1 4
    × 1.4 1 4
      5 6 5 6
    1 4 1 4 0
    5 6 5 6 0 0
  1 4 1 4 0 0 0
  1.9 9 9 3 9 6
```

27. Divide the total by the number of games. Use a calculator.

$$\frac{4621}{27} \approx 171.15$$

Drop the amount to the right of the decimal point.

171. 15

This is ——⌐ └—— Drop this amount
the average
The bowler's average is 171.

Exercise Set 8.2

1. Go down the left column (headed "Penny Number") to 20. Then go across to the column headed "Length" (immediately to the right of the Penny Number column) and read the entry, 4. The Length column gives the length in inches, so a 20-penny nail is 4 inches long.

3. Go down the left column (headed "Penny Number") to 10. Then go across to the columns headed "Approx. Number Per Pound," and find the column headed "Box Nails." The entry is 94, so there are approximately 94 10-penny box nails in a pound.

5. First, go down the "Penny Number" column to 16, and then go across to the "Finishing Nails" column to find how many 16-penny finishing nails there are in a pound. There are approximately 90. Then go down the "Penny Number" column to 10 and across to the "Common Nails" column to find how many 10-penny common nails there are in a pound. There are approximately 69. The difference, 90-69, or 21, tells us that there are approximately 21 more 16-penny finishing nails.

7. Go down the "Penny Number" column to 30, and then go across to the "Box Nail" column. There is no entry in this position. Therefore, we cannot answer this question with the information available.

9. Go down the "Penny Number" column to 4, and then go across to the "Finishing Nails" column. The entry is 548, so there are approximately 548 4-penny finishing nails in one pound. Then there are approximately 5 × 548, or 2740 nails in 5 pounds.

11. Go down the "Activity" column to Racquetball and then go across to the "154 lbs" column under the heading "Calories Burned in 30 Minutes." The entry is 294, so 294 calories are burned by a 154-pound person after 30 minutes of racquetball.

13. Go down the "110 lbs" column to 216, and then go across to the "Activity" column. The entry is Calisthenics, so calisthenics burns 216 calories in 30 minutes for a 110-pound person.

15. Going down the "Activity" column to Aerobic Dance and then across to the "154 lbs" column, we find the entry 282. Next, going down the "Activity" column to Tennis and across to the "154 lbs" column, we find the entry 222. Since 282 > 222, aerobic dance burns more calories in 30 minutes for a 154-pound person.

17. Going down the "Activity" column to Tennis and then across to the "110 lbs" column, we see that 165 calories are burned after 30 minutes of tennis for a 110-pound person. Now, 2 hours, or 120 minutes, is 4 × 30 minutes, so we multiply to find the answer: 4 × 165 calories = 660 calories.

19. Go down the "132 lbs" column and find the smallest entry. It is 132 in the bottom row. Go across the bottom row to the "Activity" column, and read the entry, Moderate Walking.

21. We observe that 120 pounds is approximately half-way between 110 pounds and 132 pounds (or is approximately the average of 110 pounds and 132 pounds). Therefore, we would expect the number of calories burned by a 120-pound person during 30 minutes of moderate walking to be approximately half-way between (or the average of) the numbers of calories burned by a 110-pound person and a 132-pound person during 30 minutes of moderate walking. Going down the "Activity" column to Moderate Walking and then across, first to the "110 lbs" column, and then to the "132 lbs" column we find the entries 111 and 132, respectively. The number half-way

21. (continued)

between these two (or their average) is $\frac{111 + 132}{2}$, or 121.5, or approximately 121. (We round down since 120 is closer to 100 than to 132.) Therefore, you would expect to burn about 121 calories during 30 minutes of moderate walking.

23. There are more bottle symbols beside 1987 than any other year. Therefore, the year in which the greatest number of bottles was sold is 1987.

25. There was positive growth between every pair of consecutive years except 1985 and 1986. The pair of years for which the amount of positive growth was the least was 1983 and 1984 (represented by an increase of 1 bottle symbol as opposed to 2 or 3 symbols for each of the other pairs).

27. Sales for 1985 are represented by 7 bottles. Since each bottle symbol stands for 1000 bottles sold, we multiply to find 7 × 1000, or 7000, bottles were sold in 1985.

29. We look for a row of the chart containing fewer symbols than the one immediately above it. The only such row is the one showing sales in 1986. Therefore, in 1986 there was a decline in the number of bottles sold.

31. The key at the bottom of the pictograph tells us that one baseball symbol represents 3 times at bat.

33. Singles are represented by 5 whole symbols (5 × 3) and 1/3 of another symbol (1/3 × 3) for a total of 15 + 1, or 16.

35. Triples are represented by 1 whole symbol, so there were 1 × 3, or 3, triples. In Exercise 11 we calculated that there were 16 singles. Then there were 16 - 3, or 13, fewer triples than singles.

37. 12 ÷ 3 = 4, so we are looking for a row containing exactly 4 whole symbols. Only the row representing outs contains exactly 4 whole symbols, so the player made an out exactly 12 times.

39. Since each license plate symbol represents 10 million cars, we divide the number of cars for each year by 10 million to determine the number of symbols to use.

    1960 - 62.5/10 = 6.25
    1965 - 75.0/10 = 7.5
    1970 - 90.0/10 = 9
    1975 - 110.0/10 = 11
    1980 - 135.0/10 = 13.5
    1985 - 157.5/10 = 15.75

Fill in the pictograph with the appropriate number of symbols beside each year.

39. (continued)

U.S. Automobile Registrations	
1960	🚗 🚗 🚗 🚗 🚗 🚗 🚗 🚗
1965	🚗 🚗 🚗 🚗 🚗 🚗 🚗 🚗
1970	🚗 🚗 🚗 🚗 🚗 🚗 🚗 🚗 🚗 🚗
1975	🚗 🚗 🚗 🚗 🚗 🚗 🚗 🚗 🚗 🚗 🚗
1980	🚗 🚗 🚗 🚗 🚗 🚗 🚗 🚗 🚗 🚗 🚗 🚗 🚗
1985	🚗 🚗 🚗 🚗 🚗 🚗 🚗 🚗 🚗 🚗 🚗 🚗 🚗 🚗 🚗 🚗

🚗 = 10 million cars

41. **Method 1:** Solve using an equation.

Restate: What is 90% of 30,955?
                ↓   ↓   ↓   ↓   ↓
Translate: a = 90% × 30,955

We multiply.

    3 0,9 5 5
  ×   0.9   (90% = 0.90 = 0.9)
  2 7,8 5 9.5

Thus, 27,859.5 sq mi of Maine are forest.

Method 2: Solve using a proportion.

$$\frac{90}{100} = \frac{a}{30,955}$$

$$90 \cdot 30,955 = 100 \cdot a$$

$$\frac{90 \cdot 30,955}{100} = a$$

$$\frac{2,785,950}{100} = a$$

$$27,859.50 = a$$

Thus, 27,859.5 sq mi of Maine are forest.

Exercise Set 8.3

1. We look for the longest bar and find that it represents Los Angeles, CA.

3. Go to the right end of the bar representing Pittsburgh, PA, and then go down to the "Weeks of Growing Season" scale. We can read, fairly accurately, that the growing season is 22 weeks long.

5. Going to the right end of the bar representing Fresno, CA, and then down to the scale, we find that the growing season in Fresno is 40 weeks long. Doing the same for the bar representing Ogden, UT, we find that the growing season in Ogden is 20 weeks long. Since 40 ÷ 20 = 2, the growing season in Fresno, CA, is 2 times as long as in Ogden, UT.

7. Going to the right end of the bar representing Los Angeles, CA, and then down to the scale, we find that the growing season in Los Angeles is 52 weeks. Now 1/2 × 52 = 26, so we go to 26 on the "Weeks of Growing Season" scale and then go up looking for the bar that ends closest to 26 weeks. It is the top bar. We then go across to the left and read the name of the city, Syracuse, NY.

9. Go to the top of the bar representing New York. Then go across to the "Daily Expenses" scale. We can read, fairly accurately, that the average daily expenses are approximately $270.

11. We look for the shortest bar and find that it represents San Francisco.

13. Going to the top of the bar representing Washington and then across to the "Daily Expenses" scale, we find that the average daily expenses are approximately $240. Doing the same for the bar representing San Francisco, we find that the average daily expenses in San Francisco are approximately $190. Subtracting, we find that the average daily expenses in Washington are $240 - $190, or approximately $50, more than in San Francisco.

15. Label the marks on the vertical scale with the names of the activities and title it "Activity." Label the horizontal scale appropriately by 100's and title it "Calories Burned in One Hour by a 152-pound Person." Then draw horizontal bars to show the calories burned by each activity.

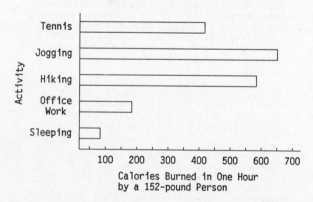

17. Find the highest point on the line, and go down to the "Time" scale. The temperature was highest at 3 P.M.

19. Finding the highest point on the line and going straight across to the "Temperature" scale, we see that the highest temperature was about 51° F. Doing the same for the lowest point on the line, we see that the lowest temperature was about 36° F. The difference is 51° F − 36° F, or approximately 15° F.

21. Reading the graph from left to right, we see that the line went down the most between 5 P.M. and 6 P.M.

23. Find the highest point on the line, and go down to the "Year" scale. Estimated sales are the greatest in 1991.

25. Find 1989 on the bottom scale and go up to the point on the line directly above it. Then go straight across to the "Estimated Sales" scale. We are at about 17.0, so estimated sales in 1989 are approximately $17.0 million.

27. The graph shows estimated sales of about $19.5 million in 1991 and about $18.0 million in 1993. Now, $19.5 million − $18.0 million = $1.5 million, so estimated sales in 1991 are approximately $1.5 million greater than in 1993.

29. First, label the horizontal scale with the times, and title it "Time (P.M.)." Then label the vertical scale by 10's, and title it "Number of Cars Counted Each Hour." Next, mark points above the times at appropriate levels, and then draw line segments connecting them.

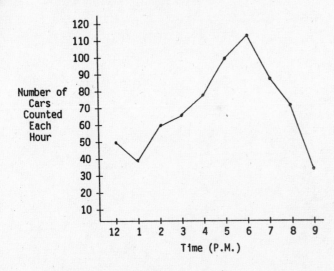

31. Reading the graph from left to right, we see that the line went down the most between 8 P.M. and 9 P.M.

33. Method 1: Solve using an equation.

Restate: 9 is what percent of 50?

Translate: 9 =   n   × 50

To solve the equation we divide on both sides by 50.

$$9 \div 50 = n$$
$$0.18 = n$$
$$18\% = n$$

Thus, 18% of the pizzas will be ordered with extra cheese.

Method 2: Solve using a proportion.

$$\frac{n}{100} = \frac{9}{50}$$
$$50 \cdot n = 100 \cdot 9$$
$$n = \frac{100 \cdot 9}{50}$$
$$n = \frac{900}{50}$$
$$n = 18$$

Thus, 18% of the pizzas will be ordered with extra cheese.

35. 34 is what percent of 51?

34 =   n   × 51

To solve the equation, we divide on both sides by 51.

$$34 \div 51 = n$$
$$0.66\overline{6} = n$$
$$66.\overline{6}\% = n, \text{ or}$$
$$66\tfrac{2}{3}\% = n$$

34 is 66.$\overline{6}$%, or 66$\tfrac{2}{3}$%, of 51.

Exercise Set 8.4

1. We see from the graph that 3.7% of all records sold are jazz.

3. We see from the graph that 9% of all records sold are country. Find 9% of 3000: 0.09 × 3000 = 270. Then 270 of the records are country.

5. We see from the graph that 6.8% of all records sold are classical.

7. The largest section of the circle represents gas purchased. Therefore, the most is spent on gas purchased.

9. We see from the graph that 4¢ of each dollar is spent on dividends.

<u>11.</u> 4¢ of each dollar is spent on dividends, and 12¢
    on wages, salaries, and employee benefits.
    Adding these amounts, we find that 4¢ + 12¢, or
    16¢, of every dollar is spent on these items all
    together.

<u>13.</u> First find the number of degrees required to
    represent each type of expenditure.
    Transportation accounts for 15% of the
    expenditures. 15% of 360° (0.15 × 360°) is 54°.
    Meals account for 20% of the expenditures.
    20% of 360° (0.2 × 360°) is 72°. Lodging accounts
    for 32% of the expenditures. 32% of 360°
    (0.32 × 360°) is about 115°. Recreation accounts
    for 18% of the expenditures. 18% of 360°
    (0.18 × 360°) is about 65°. Other accounts for
    15% of the expenditures. As computed above,
    15% of 360° is 54°. Using this information we
    can draw a circle graph and label it
    appropriately.

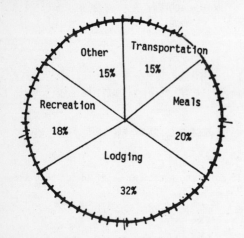

<u>15.</u> 16 is what percent of 64?
     ↓ ↓   ↓   ↓ ↓
    16 =   n   × 64

    To solve the equation we divide on both sides
    by 64.

        16 ÷ 64 = n
          0.25 = n
           25% = n
    16 is 25% of 64.

<u>17.</u> $\frac{4}{7} \div \frac{16}{21} = \frac{4}{7} \cdot \frac{21}{16} = \frac{4 \cdot 21}{7 \cdot 16} = \frac{4 \cdot 3 \cdot 7}{7 \cdot 4 \cdot 4} = \frac{4 \cdot 7}{4 \cdot 7} \cdot \frac{3}{4} = \frac{3}{4}$

Exercise Set 9.1

1. 1 foot = 12 in.

    This is the relation stated on page 383 of the text.

3. 1 in. = 1 in. $\times \frac{1 \text{ ft}}{12 \text{ in.}}$ 1 ft = 12 in., so $\frac{1 \text{ ft}}{12 \text{ in.}}$ = 1. We use $\frac{1 \text{ ft}}{12 \text{ in.}}$ to eliminate in.

    $= \frac{1}{12} \times \frac{\text{in.}}{\text{in.}} \times 1 \text{ ft}$

    $= \frac{1}{12} \times 1 \text{ ft}$ The $\frac{\text{in.}}{\text{in.}}$ acts like 1, so we can omit it.

    $= \frac{1}{12} \text{ ft}$

5. 1 mi = 5280 ft

    This is the relation stated on page 383 of the text.

7. 13 yd = 13 × 1 yd

    = 13 × 3 ft   Substituting 3 ft for 1 yd

    = 13 × 3 × 1 ft

    = 13 × 3 × 12 in.   Substituting 12 in. for 1 ft

    = 468 in.   Multiplying

9. 84 in. = $\frac{84 \text{ in.}}{1} \times \frac{1 \text{ ft}}{12 \text{ in.}}$ Multiplying by 1 using $\frac{1 \text{ ft}}{12 \text{ in.}}$

    $= \frac{84 \text{ in.}}{12 \text{ in.}} \times 1 \text{ ft}$

    $= \frac{84}{12} \times \frac{\text{in.}}{\text{in.}} \times 1 \text{ ft}$

    = 7 × 1 ft   The $\frac{\text{in.}}{\text{in.}}$ acts like 1,

    = 7 ft   so we can omit it.

11. 18 in. = $\frac{18 \text{ in.}}{1} \times \frac{1 \text{ ft}}{12 \text{ in.}}$ Multiplying by 1 using $\frac{1 \text{ ft}}{12 \text{ in.}}$

    $= \frac{18 \text{ in.}}{12 \text{ in.}} \times 1 \text{ ft}$

    $= \frac{18}{12} \times \frac{\text{in.}}{\text{in.}} \times 1 \text{ ft}$

    $= \frac{3}{2} \times 1 \text{ ft}$ The $\frac{\text{in.}}{\text{in.}}$ acts like 1, so we can omit it.

    $= \frac{3}{2} \text{ ft or } 1 \frac{1}{2} \text{ ft}$

13. 3 mi = 3 × 1 mi

    = 3 × 5280 ft   Substituting 5280 ft for 1 mi

    = 15,840 ft   Multiplying

15. 3 in. = $\frac{3 \text{ in.}}{1} \times \frac{1 \text{ ft}}{12 \text{ in.}}$ Multiplying by 1 using $\frac{1 \text{ ft}}{12 \text{ in.}}$

    $= \frac{3 \text{ in.}}{12 \text{ in.}} \times 1 \text{ ft}$

    $= \frac{3}{12} \times \frac{\text{in.}}{\text{in.}} \times 1 \text{ ft}$

    $= \frac{1}{4} \times 1 \text{ ft}$ The $\frac{\text{in.}}{\text{in.}}$ acts like 1, so we can omit it.

    $= \frac{1}{4} \text{ ft}$

17. 10 ft = 10 ft $\times \frac{1 \text{ yd}}{3 \text{ ft}}$ Multiplying by 1 using $\frac{1 \text{ yd}}{3 \text{ ft}}$

    $= \frac{10}{3} \times \frac{\text{ft}}{\text{ft}} \times 1 \text{ yd}$

    $= \frac{10}{3} \times 1 \text{ yd}$

    $= \frac{10}{3} \text{ yd or } 3 \frac{1}{3} \text{ yd}$

19. 10 mi = 10 × 1 mi

    = 10 × 5280 ft   Substituting 5280 ft for 1 mi

    = 52,800 ft   Multiplying

21. $4\frac{1}{2}$ ft = $4\frac{1}{2}$ ft $\times \frac{1 \text{ yd}}{3 \text{ ft}}$

    $= \frac{9}{2} \text{ ft} \times \frac{1 \text{ yd}}{3 \text{ ft}}$

    $= \frac{9}{6} \times \frac{\text{ft}}{\text{ft}} \times 1 \text{ yd}$

    $= \frac{3}{2} \times 1 \text{ yd}$

    $= \frac{3}{2} \text{ yd or } 1\frac{1}{2} \text{ yd}$

23. 36 in. = 36 in. $\times \frac{1 \text{ ft}}{12 \text{ in.}}$

    = 36 in. $\times \frac{1 \text{ ft}}{12 \text{ in.}} = \frac{36}{12} \times \frac{\text{in.}}{\text{in.}} \times 1 \text{ ft}$

    = 3 × 1 ft = 3 ft

    = 3 ft $\times \frac{1 \text{ yd}}{3 \text{ ft}} = \frac{3}{3} \times \frac{\text{ft}}{\text{ft}} \times 1 \text{ yd}$

    = 1 × 1 yd

    = 1 yd

25. 330 ft = 330 ft $\times \frac{1 \text{ yd}}{3 \text{ ft}}$

    $= \frac{330}{3} \times \frac{\text{ft}}{\text{ft}} \times 1 \text{ yd}$

    = 110 × 1 yd

    = 110 yd

27. 3520 yd = 3520 × 1 yd = 3520 × 3 ft

    = 10,560 ft = 10,560 ft $\times \frac{1 \text{ mi}}{5280 \text{ ft}}$

    $= \frac{10,560}{5280} \times \frac{\text{ft}}{\text{ft}} \times 1 \text{ mi} = 2 \times 1 \text{ mi}$

    = 2 mi

<u>29.</u> 100 yd = 100 × 1 yd
          = 100 × 3 ft
          = 300 ft

<u>31.</u> 360 in. = 360 in. × $\frac{1 \text{ ft}}{12 \text{ in.}}$

        = $\frac{360}{12}$ × $\frac{\text{in.}}{\text{in.}}$ × 1 ft

        = 30 × 1 ft

        = 30 ft

<u>33.</u> 1 in. = 1 in. × $\frac{1 \text{ ft}}{12 \text{ in.}}$

      = $\frac{1}{12}$ × $\frac{\text{in.}}{\text{in.}}$ × 1 ft = $\frac{1}{12}$ × 1 ft

      = $\frac{1}{12}$ ft = $\frac{1}{12}$ ft × $\frac{1 \text{ yd}}{3 \text{ ft}}$

      = $\frac{1}{36}$ × $\frac{\text{ft}}{\text{ft}}$ × 1 yd = $\frac{1}{36}$ × 1 yd

      = $\frac{1}{36}$ yd

<u>35.</u> 2 mi = 2 × 1 mi = 2 × 5280 ft = 10,560 ft
      = 10,560 × 1 ft = 10,560 × 12 in.
      = 126,720 in.

<u>37.</u> 23.87 × <u>10</u>   23.8.7

    1 zero   Move 1 place

   23.87 × 10 = 238.7

<u>39.</u> 23.87 × <u>1000</u>   23.870.

    3 zeros   Move 3 places

   23.87 × 1000 = 23,870

<u>41.</u> First, convert 119,433 mi to inches. Then divide by 1.824 trillion.

     119,433 mi = 119,433 × 1 mi
              = 119,433 × 5280 ft
              = 630,606,240 ft
              = 630,606,240 × 1 ft
              = 630,606,240 × 12 in.
              = 7,567,274,880 in.

7,567,274,880 ÷ 1,824,000,000,000 ≈ 0.0041
A $1 bill is approximately 0.0041 in. thick.

Exercise Set 9.2

<u>1.</u> a) 1 km = _____ m

    Think: To go from km to m in the table is a move of 3 places to the right. Thus we move the decimal point 3 places to the right.

       1   1.000.

    1 km = 1000 m

  b) 1 m = _____ km

    Think: To go from m to km in the table is a move of 3 places to the left. Thus we move the decimal point 3 places to the left.

       1   .001.

    1 m = 0.001 km

<u>3.</u> a) 1 dam = _____ m

    Think: To go from dam to m in the table is a move of 1 place to the right. Thus we move the decimal point 1 place to the right.

       1   1.0.

    1 dam = 10 m

  b) 1 m = _____ dam

    Think: To go from m to dam in the table is a move of 1 place to the left. Thus we move the decimal point 1 place to the left.

       1   .1.

    1 m = 0.1 dam

<u>5.</u> a) 1 cm = _____ m

    Think: To go from cm to m in the table is a move of 2 places to the left. Thus we move the decimal point 2 places to the left.

       1   .01.

    1 cm = 0.01 m

  b) 1 m = _____ cm

    Think: To go from m to cm in the table is a move of 2 places to the right. Thus we move the decimal point 2 places to the right.

       1   1.00.

    1 m = 100 cm

7. 6.7 km = _____ m

Think: To go from km to m in the table is a move of 3 places to the right. Thus we move the decimal point 3 places to the right.

    6.7   6.700.

6.7 km = 6700 m

9. 98 cm = _____ m

Think: To go from cm to m in the table is a move of 2 places to the left. Thus we move the decimal point 2 places to the left.

    98   0.98.

98 cm = 0.98 m

11. 8921 m = _____ km

Think: To go from m to km in the table is a move of 3 places to the left. Thus we move the decimal point 3 places to the left.

    8921   8.921.

8921 m = 8.921 km

13. 56.66 m = _____ km

Think: To go from m to km in the table is a move of 3 places to the left. Thus we move the decimal point 3 places to the left.

    56.66   0.056.66

56.66 m = 0.05666 km

15. 5666 m = _____ cm

Think: To go from m to cm in the table is a move of 2 places to the right. Thus we move the decimal point 2 places to the right.

    5666   5666.00.

5666 m = 566,600 cm

17. 477 cm = _____ m

Think: To go from cm to m in the table is a move of 2 places to the left. Thus we move the decimal point 2 places to the left.

    477   4.77.

477 cm = 4.77 m

19. 6.88 m = _____ cm

Think: To go from m to cm in the table is a move of 2 places to the right. Thus we move the decimal point 2 places to the right.

    6.88   6.88.

6.88 m = 688 cm

21. 1 mm = _____ cm

Think: To go from mm to cm in the table is a move of 1 place to the left. Thus we move the decimal point 1 place to the left.

    1   0.1.

1 mm = 0.1 cm

23. 1 km = _____ cm

Think: To go from km to cm in the table is a move of 5 places to the right. Thus we move the decimal point 5 places to the right.

    1   1.00000.

1 km = 100,000 cm

25. 14.2 cm = _____ mm

Think: To go from cm to mm in the table is a move of 1 place to the right. Thus we move the decimal point 1 place to the right.

    14.2   14.2.

14.2 cm = 142 mm

27. 8.2 mm = _____ cm

Think: To go from mm to cm in the table is a move of 1 place to the left. Thus we move the decimal point 1 place to the left.

    8.2   0.8.2

8.2 mm = 0.82 cm

29. 4500 mm = _____ cm

Think: To go from mm to cm in the table is a move of 1 place to the left. Thus we move the decimal point 1 place to the left.

    4500   450.0.

4500 mm = 450 cm

31. 0.024 mm = _____ m

Think: To go from mm to m in the table is a move of 3 places to the left. Thus we move the decimal point 3 places to the left.

    0.024   0.000.024

    0.024 mm = 0.000024 m

33. 6.88 m = _____ dam

Think: To go from m to dam in the table is a move of 1 place to the left. Thus we move the decimal point 1 place to the left.

    6.88   0.6.88

    6.88 m = 0.688 dam

35. 2.3 dam = _____ dm

Think: To go from dam to dm in the table is a move of 2 places to the right. Thus we move the decimal point 2 places to the right.

    2.3   2.30.

    2.3 dam = 230 dm

37. 392 dam = _____ km

Think: To go from dam to km in the table is a move of 2 places to the left. Thus we move the decimal point 2 places to the left.

    392   3.92.

    392 dam = 3.92 km

39. $330 \text{ ft} \approx 330 \text{ ft} \times \dfrac{1 \text{ m}}{3.3 \text{ ft}} = \dfrac{330}{3.3} \times \dfrac{\text{ft}}{\text{ft}} \times 1 \text{ m} = 100 \text{ m}$

41. $2 \text{ m} = 2 \times 1 \text{ m} = 2 \times 3.3 \text{ ft} = 6.6 \text{ ft}$

43. $55 \text{ mph} = 55 \dfrac{\text{mi}}{\text{hr}} = 55 \times \dfrac{1 \text{ mi}}{\text{hr}}$

    $55 \times \dfrac{1.609 \text{ km}}{\text{hr}} = 88.495 \dfrac{\text{km}}{\text{hr}}$

45.
```
        1.7 5
  1 2 ) 2 1.0 0
        1 2
          9 0
          8 4
            6 0
            6 0
               0
```

47. To divide by 100, move the decimal point 2 places to the left.

    23.4   .23.4

    23.4 ÷ 100 = 0.234

49.
```
      3.1 4   (2 decimal places)
      4.4 1   (2 decimal places)
      3 1 4
    1 2 5 6 0
  1 2 5 6 0 0
  1 3.8 4 7 4   (4 decimal places)
```

Round to the nearest hundredth:

    13.84 7 4   Thousandths digit is 5 or
                     higher. Round up.
    13.85

51. $48 \cdot \dfrac{1}{12} = \dfrac{48 \cdot 1}{12} = \dfrac{48}{12} = 4$

Exercise Set 9.3

1. Perimeter = 4 mm + 6 mm + 7 mm
             = (4 + 6 + 7) mm
             = 17 mm

3. Perimeter = 3.5 cm + 3.5 cm + 4.25 cm + 0.5 cm + 3.5 cm
             = (3.5 + 3.5 + 4.25 + 0.5 + 3.5) cm
             = 15.25 cm

5. P = 4·s   Perimeter of a square
   P = 4·3.25 m
     = 13 m

7. P = 2·(ℓ + w)   Perimeter of a rectangle
   P = 2·(5 ft + 10 ft)
   P = 2·(15 ft)
   P = 30 ft

9. P = 2·(ℓ + w)   Perimeter of a rectangle
   P = 2·(34.67 cm + 4.9 cm)
   P = 2·(39.57 cm)
   P = 79.14 cm

11. P = 4·s   Perimeter of a square
    P = 4·22 ft
    P = 88 ft

13. P = 4·s   Perimeter of a square
    P = 4·45.5 mm
    P = 182 mm

15. Familiarize. First we will find the perimeter of the field. Then we will multiply to find the cost of the fence wire. We make a drawing.

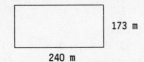

173 m

240 m

Translate. The perimeter of the field is given by
P = 2·(ℓ + w) = 2·(240 m + 173 m).

Solve. We calculate the perimeter.
P = 2·(240 m + 173 m) = 2·(413 m) = 826 m

Then we multiply to find the cost of the fence wire.
Cost = $1.45/m × Perimeter =
$1.45/m × 826 m = $1197.70

Check. Repeat the calculations.

State. The perimeter of the field is 826 m. The fence wire will cost $1197.70.

17. Familiarize. We make a drawing.

27.9 cm

21.6 cm

Translate. The perimeter of the sheet of paper is given by
P = 2·(ℓ + w) = 2(27.9 cm + 21.6 cm)

Solve. We do the calculation.
2(27.9 cm + 21.6 cm) = 2·(49.5 cm) = 99 cm

Check. We repeat the calculation.

State. The perimeter is 99 cm.

19. Familiarize. We first find the perimeter of the garden. We make a drawing.

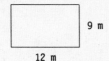

9 m

12 m

Translate. The perimeter is given by
P = 2·(ℓ + w) = 2·(12 m + 9 m).

Solve. We do the calculation.
P = 2·(12 m + 9 m) = 2·(21 m) = 42 m

a) We divide to find n, the number of posts needed.
n = 42 ÷ 3 = 14
Thus, 14 posts are needed.

b) We multiply to find the cost of the posts.
Cost = $2.40 × 14 = $33.60
The posts will cost $33.60.

19. (continued)

c) We subtract to find the length F of the fence.
F = 42 m − 3 m = (42 − 3) m = 39 m
The fence will be 39 m long.

d) We multiply to find the cost of the fence.
Cost = $0.85/m × 39 m = $33.15
The fence will cost $33.15.

e) We add to find the total cost of the materials.
Total cost = Cost of posts +
Cost of fence + Cost of gate =
$33.60 + $33.15 + $9.95 = $76.70

Check. Repeat the calculations.

State. See the results stated in (a) - (e) above.

21. 56.1%

a) Drop the percent symbol.   56.1

b) Move the decimal point   0.56.1
two places to the left.

56.1% = 0.561

23. $31^2 = 31 \times 31 = 961$

___

Exercise Set 9.4

1. A = ℓ·w   Area of a rectangular region
A = (5 km)·(3 km)
= 5·3·km·km
= 15 km²

3. A = ℓ·w   Area of a rectangular region
A = (2 cm)·(0.7 cm)
= 2·0.7·cm·cm
= 1.4 cm²

5. A = s·s   Area of a square
A = (2.5 mm)·(2.5 mm)
= 2.5·2.5·mm·mm
= 6.25 mm²

7. A = s·s   Area of a square
A = (90 ft)·(90 ft)
= 90·90·ft·ft
= 8100 ft²

9. A = ℓ·w
A = (10 ft)·(5 ft)
= 10·5·ft·ft
= 50 ft²

11. $A = l \cdot w$

    $A = (34.67 \text{ cm}) \cdot (4.9 \text{ cm})$

      $= 34.67 \cdot 4.9 \cdot \text{cm} \cdot \text{cm}$

      $= 169.883 \text{ cm}^2$

13. $A = s \cdot s$

    $A = (22 \text{ ft}) \cdot (22 \text{ ft})$

      $= 22 \cdot 22 \cdot \text{ft} \cdot \text{ft}$

      $= 484 \text{ ft}^2$

15. $A = s \cdot s$

    $A = (56.9 \text{ km}) \cdot (56.9 \text{ km})$

      $= 56.9 \cdot 56.9 \cdot \text{km} \cdot \text{km}$

      $= 3237.61 \text{ km}^2$

17. <u>Familiarize</u>. We draw a picture.

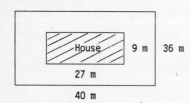

   <u>Translate</u>. We let A = the area left over.

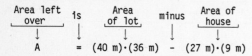

   <u>Solve</u>. The area of the lot is

     $(40 \text{ m}) \cdot (36 \text{ m}) = 40 \cdot 36 \cdot \text{m} \cdot \text{m} = 1440 \text{ m}^2$.

   The area of the house is

     $(27 \text{ m}) \cdot (9 \text{ m}) = 27 \cdot 9 \cdot \text{m} \cdot \text{m} = 243 \text{ m}^2$.

   The area left over is

     $A = 1440 \text{ m}^2 - 243 \text{ m}^2 = 1197 \text{ m}^2$.

   <u>Check</u>. Repeat the calculations.

   <u>State</u>. The area left over is 1197 m².

19. <u>Familiarize</u>. We use the drawing in the text.

   <u>Translate</u>. We let A = the area of the sidewalk.

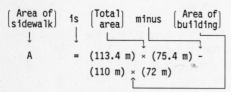

   <u>Solve</u>. The total area is

   $(113.4 \text{ m}) \times (75.4 \text{ m}) = 113.4 \times 75.4 \times \text{m} \times \text{m} =$

   $= 8550.36 \text{ m}^2$.

19. (continued)

   The area of the building is

   $(110 \text{ m}) \times (72 \text{ m}) = 110 \times 72 \times \text{m} \times \text{m} = 7920 \text{ m}^2$.

   The area of the sidewalk is

   $A = 8550.36 \text{ m}^2 - 7920 \text{ m}^2 = 630.36 \text{ m}^2$.

   <u>Check</u>. Repeat the calculations.

   <u>State</u>. The area of the sidewalk is 630.36 m².

21. <u>Familiarize</u>. First we will find the area of the room. Then we will find the cost of the carpeting.

   <u>Translate</u>. To find how many square meters of carpeting will be needed, find the area of the room.

       $A = l \cdot w = (5.5 \text{ m}) \times (4.5 \text{ m})$

   <u>Solve</u>. a) We find the area.

     $A = (5.5 \text{ m}) \times (4.5 \text{ m}) = 5.5 \times 4.5 \times \text{m} \times \text{m}$

      $= 24.75 \text{ m}^2$

   Thus, 24.75 m² of carpeting will be needed.

   b) We multiply to find the cost of the carpet.

      Cost = \$8.40/m² × 24.75 m² = \$207.90

   The carpet will cost \$207.90.

   <u>Check</u>. Repeat the calculations.

   <u>State</u>. See the results stated in (a) and (b) above.

23. Think of this figure as a large rectangular region containing a smaller rectangular region. The area of the shaded region is the area left over when the area of the smaller region is taken away from the area of the larger region.

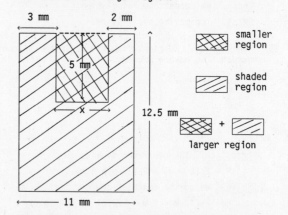

The larger rectangle measures 12.5 mm by 11 mm. One dimension of the smaller rectangle is 5 mm. Let x represent the other dimension. To find x we subtract:

   $x = 11 \text{ m} - 3 \text{ m} - 2 \text{ m} = (11 - 3 - 2) \text{ mm} = 6 \text{ mm}$

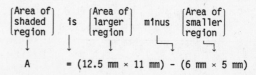

23. (continued)

    The area of the larger region is
    (12.5 mm) × (11 mm) = 12.5 × 11 × mm × mm = 137.5 mm².

    The area of the smaller region is
    (6 mm) × (5 mm) = 6 × 5 × mm × mm = 30 mm².

    The area of the shaded region is
    A = 137.5 mm² - 30 mm² = 107.5 mm².

25. 0.452   0.45.2   45.2%

    Move decimal   Write
    point 2 places   a %
    to the right   symbol

    0.452 = 45.2%

27. We multiply by 1 to get 100 in the denominator.

    $\frac{11}{20} = \frac{11}{20} \cdot \frac{5}{5} = \frac{55}{100} = 55\%$

Exercise Set 9.5

1. A = b·h   Area of a parallelogram
   A = 8 cm·4 cm   Substituting 8 cm for b and
                          4 cm for h
     = 32 cm²

3. $A = \frac{1}{2}·b·h$ Area of a triangle

   $A = \frac{1}{2}·12$ m·6 m   Substituting 12 m for b and
                          6 m for h

   $= \frac{12·6}{2} \cdot$ m²

   = 36 m²

5. $A = \frac{1}{2}·h·(a + b)$ Area of a trapezoid

   $A = \frac{1}{2}·6$ ft·(5 + 12) ft   Substituting 6 ft for
                          h, 5 ft for a, and 12
                          ft for b

   $= \frac{6·17}{2}$ ft² = 51 ft²

7. A = b·h   Area of a parallelogram
   A = 8 m·8 m
     = 64 m²

9. $A = \frac{1}{2}·h·(a + b)$ Area of a trapezoid

   $A = \frac{1}{2}·7$ mm·(4.5 mm + 8.5 mm)   Substituting 7 mm
                          for h, 4.5 mm for
                          a, and 8.5 mm for
                          b

   $= \frac{7·13}{2} \cdot$ mm²

   $= \frac{91}{2}$ mm²

   = 45.5 mm²

11. A = b·h   Area of a parallelogram
    A = 2.3 cm·3.5 cm   Substituting 2.3 cm for b and
                          3.5 cm for h
      = 8.05 cm²

13. $A = \frac{1}{2}·h·(a + b)$ Area of a trapezoid

    $A = \frac{1}{2}·18$ cm·(9 + 24) cm   Substituting 18 cm
                          for h, 9 cm for a,
                          and 24 cm for b

    $= \frac{18·33}{2}$ cm²

    = 297 cm²

15. $A = \frac{1}{2}·b·h$ Area of a triangle

    $A = \frac{1}{2}·4$ m·3.5 m   Substituting 4 m for b and
                          3.5 m for h

    $= \frac{4·3.5}{2}$ m²

    = 7 m²

17. A = b·h   Area of a parallelogram

    $A = 12 \frac{1}{4}$ ft·$4 \frac{1}{2}$ ft   Substituting $12 \frac{1}{4}$ ft for b
                          and $4 \frac{1}{2}$ ft for h

    $= \frac{49·9}{4·2}$·ft

    $= \frac{441}{8}$ ft²

    $= 55 \frac{1}{8}$ ft²

19. Familiarize. We look for the kinds of figures whose areas we can calculate using area formulas that we already know.

    Translate. The shaded region consists of a square region with a triangular region removed from it. The sides of the square are 30 cm, and the triangle has base 30 cm and height 15 cm. We find the area of the square using the formula A = s·s and the area of the triangle using $A = \frac{1}{2}·b·h$. Then we subtract.

    Solve.
        Area of the square:
            A = 30 cm·30 cm = 900 cm²
        Area of the triangle:
            $A = \frac{1}{2} \cdot$ 30 cm·15 cm = 225 cm²
        Area of the shaded region:
            A = 900 cm² - 225 cm² = 675 cm²
    Check. We repeat the calculations.
    State. The area of the shaded region is 675 cm².

<u>21.</u> <u>Familiarize</u>. We look for the kinds of figures whose areas we can calculate using area formulas that we already know.

<u>Translate</u>. The shaded region consists of 8 triangles, each with base 52 in. and height 52 in. We will find the area of one triangle using the formula $A = \frac{1}{2} \cdot b \cdot h$. Then we will multiply by 8.

<u>Solve</u>.

$A = \frac{1}{2} \cdot 52$ in. $\cdot 52$ in. $= 1352$ in²

Then we multiply by 8:

$8 \cdot 1352$ in² $= 10,816$ in²

<u>Check</u>. We repeat the calculations.

<u>State</u>. The area of the shaded region is 10,816 in².

<u>23.</u> <u>Familiarize</u>. We make a drawing, shading the area left over after the pool is constructed.

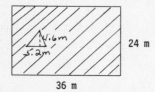

<u>Translate</u>. The shaded region consists of a rectangular region with a triangular region removed from it. The rectangular region has dimensions 36 m by 24 m, and the triangular region has base 5.2 m and height 4.6 m. We wil find the area of the rectangular region using the formula $A = b \cdot h$ and the area of the triangular region using $A = \frac{1}{2} \cdot b \cdot h$. Then we will subtract to find the area of the shaded region.

<u>Solve</u>.

Area of rectangle:

$A = 36$ m $\cdot 24$ m $= 864$ m²

Area of triangle:

$A = \frac{1}{2} \cdot (5.2$ m$) \cdot (4.6$ m$) = 11.96$ m²

Area of shaded region:

$A = 864$ m² $- 11.96$ m² $= 852.04$ m²

<u>Check</u>. We repeat the calculations.

<u>State</u>. The area left over is 852.04 m².

<u>25.</u> $9.25\% = \frac{9.25}{100}$ Definition of percent

$ = \frac{9.25}{100} \cdot \frac{100}{100}$ Multiplying by 1 to get rid of the decimal point in the numerator

$ = \frac{925}{10,000}$

$ = \frac{37}{400} \cdot \frac{25}{25}$ Simplifying

$ = \frac{37}{400}$

<u>27.</u>

$$\begin{array}{r} 1.3\,7\,5 \\ 8\overline{)1\,1.0\,0\,0} \\ \underline{8} \\ 3\,0 \\ \underline{2\,4} \\ 6\,0 \\ \underline{5\,6} \\ 4\,0 \\ \underline{4\,0} \\ 0 \end{array}$$

$\frac{11}{8} = 1.375$

a) Move the decimal point 2 places to the right.   1.37.5

b) Add a percent symbol.   137.5%

$\frac{11}{8} = 137.5\%$

Exercise Set 9.6

<u>1.</u> $d = 2 \cdot r$

$d = 2 \cdot 7$ cm $= 14$ cm

<u>3.</u> $d = 2 \cdot r$

$d = 2 \cdot \frac{3}{4}$ in. $= \frac{6}{4}$ in. $= \frac{3}{2}$ in., or $1\frac{1}{2}$ in.

<u>5.</u> $r = \frac{d}{2}$

$r = \frac{32 \text{ ft}}{2} = 16$ ft

<u>7.</u> $r = \frac{d}{2}$

$r = \frac{1.4 \text{ cm}}{2} = 0.7$ cm

<u>9.</u> $C = 2 \cdot \pi \cdot r$

$C \approx 2 \cdot \frac{22}{7} \cdot 7$ cm $= \frac{2 \cdot 22 \cdot 7}{7}$ cm $= 44$ cm

<u>11.</u> $C = 2 \cdot \pi \cdot r$

$C \approx 2 \cdot \frac{22}{7} \cdot \frac{3}{4}$ in. $= \frac{2 \cdot 22 \cdot 3}{7 \cdot 4}$ in. $= \frac{132}{28}$ in.

$ = \frac{33}{7}$ in., or $4\frac{5}{7}$ in.

<u>13.</u> $C = \pi \cdot d$

$C \approx 3.14 \cdot 32$ ft $= 100.48$ ft

<u>15.</u> $C = \pi \cdot d$

$C \approx 3.14 \cdot 1.4$ cm $= 4.396$ cm

<u>17.</u> $A = \pi \cdot r \cdot r$

$A \approx \frac{22}{7} \cdot 7$ cm $\cdot 7$ cm $= \frac{22}{7} \cdot 49$ cm² $= 154$ cm²

19. A = π·r·r

$A \approx \frac{22}{7} \cdot \frac{3}{4}$ in. $\cdot \frac{3}{4}$ in. $= \frac{22 \cdot 3 \cdot 3}{7 \cdot 4 \cdot 4}$ in²

$= \frac{99}{56}$ in², or $1\frac{43}{56}$ in²

21. A = π·r·r

$A \approx 3.14 \cdot 16$ ft·16 ft $\left[r = \frac{d}{2};\ r = \frac{32\ ft}{2} = 16\ ft\right]$

A = 3.14·256 ft²

A = 803.84 ft²

23. A = π·r·r

$A \approx 3.14 \cdot 0.7$ cm·0.7 cm $\left[r = \frac{d}{2};\ r = \frac{1.4\ cm}{2}\right.$

$\left. = 0.7\ cm\right]$

A = 3.14·0.49 cm² = 1.5386 cm²

25. $r = \frac{d}{2}$

$r = \frac{6\ cm}{2} = 3$ cm

The radius is 3 cm.

C = π·d

C ≈ 3.14·6 cm = 18.84 cm

The circumference is about 18.84 cm.

A = π·r·r

A ≈ 3.14·3 cm·3 cm = 28.26 cm²

The area is about 28.26 cm².

27. A = π·r·r

A ≈ 3.14·220 km·220 km = 151,976 km²

The broadcast area is about 151,976 km².

29. C = π·d

C ≈ 3.14·1.1 m = 3.454 m

The circumference of the elm tree is about
3.454 m.

31. C = π·d

7.85 cm ≈ 3.14·d   Substituting 7.85 cm for C and
                      3.14 for π

$\frac{7.85\ cm}{3.14} = d$ Dividing on both sides by 3.14

2.5 cm = d

The diameter is about 2.5 cm.

$r = \frac{d}{2}$

$r = \frac{2.5\ cm}{2} = 1.25$ cm

The radius is about 1.25 cm.

A = π·r·r

A ≈ 3.14·1.25 cm·1.25 cm = 4.90625 cm²

The area is about 4.90625 cm².

33.

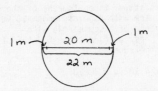

Find the area of the larger circle (pool plus
wall). Its diameter is 1 m + 20 m + 1 m, or
22 m. Thus, its radius is $\frac{22}{2}$ m or 11 m.

A = π·r·r

A ≈ 3.14·11 m·11 m = 379.94 m²

Find the area of the pool. Its diameter is 20 m.
Thus, its radius is $\frac{20}{2}$ m, or 10 m.

A = π·r·r

A ≈ 3.14·10 m·10 m = 314 m²

We subtract to find the area of the walk:

A = 379.94 m² - 314 m²

A = 65.94 m²

The area of the walk is 65.94 m².

35. The perimeter consists of the circumferences of
three semicircles, each with diameter 8 ft, and
one side of a square of length 8 ft. We first
find the circumference of one semicircle. This
is one-half the circumference of a circle with
diameter 8 ft:

$\frac{1}{2} \cdot \pi \cdot d \approx \frac{1}{2} \cdot 3.14 \cdot 8$ ft = 12.56 ft

Then we multiply by 3:

3·(12.56 ft) = 37.68 ft

Finally we add the circumferences of the
semicircles and the length of the side of the
square:

37.68 ft + 8 ft = 45.68 ft

The perimeter is 45.68 ft.

37. The perimeter consists of three-fourths of the
circumference of a circle with radius 4 yd and
two sides of a square with sides of length 4 yd.
We first find three-fourths of the circumference
of the circle:

$\frac{3}{4} \cdot 2 \cdot \pi \cdot r \approx 0.75 \cdot 2 \cdot 3.14 \cdot 4$ yd = 18.84 yd

Then we add this length to the lengths of two
sides of the square:

18.84 yd + 4 yd + 4 yd = 26.84 yd

The perimeter is 26.84 yd.

39. The perimeter consists of three-fourths of the perimeter of a square with side of length 10 yd and the circumference of a semicircle with diameter 10 yd. First we find three-fourths of the perimeter of the square:

$$\frac{3}{4} \cdot 4 \cdot s = \frac{3}{4} \cdot 4 \cdot 10 \text{ yd} = 30 \text{ yd}$$

Then we find one-half of the circumference of a circle with diameter 10 yd:

$$\frac{1}{2} \cdot \pi \cdot d \approx \frac{1}{2} \cdot 3.14 \cdot 10 \text{ yd} = 15.7 \text{ yd}$$

Then we add:

30 yd + 15.7 yd = 45.7 yd

The perimeter is 45.7 yd.

41. The shaded region consists of a circle, with radius 8 m, with two circles, each with diamter 8 m, removed. First we find the area of the large circle:

$$A = \pi \cdot r \cdot r \approx 3.14 \cdot 8 \text{ m} \cdot 8 \text{ m} = 200.96 \text{ m}^2$$

Then find the area of one of the small circles:

The radius is $\frac{8 \text{ m}}{2} = 4$ m.

$$A = \pi \cdot r \cdot r \approx 3.14 \cdot 4 \text{ m} \cdot 4 \text{ m} = 50.24 \text{ m}^2$$

We multiply this area by 2 to find the area of the two small circles:

$$2 \cdot 50.24 \text{ m}^2 = 100.48 \text{ m}^2$$

Finally we subtract to find the area of the shaded region:

$$200.96 \text{ m}^2 - 100.48 \text{ m}^2 = 100.48 \text{ m}^2$$

The area of the shaded region is 100.48 m².

43. The shaded region consists of one-half of a circle with diameter 2.8 cm and a triangle with base 2.8 cm and height 2.8 cm. First we find the area of the semicircle. The radius is $\frac{2.8 \text{ cm}}{2}$ = 1.4 cm.

$$\frac{1}{2} \cdot \pi \cdot r \cdot r \approx \frac{1}{2} \cdot 3.14 \cdot 1.4 \text{ cm} \cdot 1.4 \text{ cm} = 3.0772 \text{ cm}^2$$

Then we find the area of the triangle.

$$\frac{1}{2} \cdot b \cdot h = \frac{1}{2} \cdot 2.8 \text{ cm} \cdot 2.8 \text{ cm} = 3.92 \text{ cm}^2$$

Finally we add to find the area of the shaded region.

$$3.0772 \text{ cm}^2 + 3.92 \text{ cm}^2 = 6.9972 \text{ cm}^2$$

The area of the shaded region is 6.9972 cm².

45. The shaded region consists of a rectangle, with dimensions 10.2 cm by 12.8 cm, with the area of two semicircles, each with diameter 10.2 cm, removed. This is equivalent to removing one circle with diameter 10.2 cm from the rectangle. First we find the area of the rectangle.

$$\ell \cdot w = (10.2 \text{ cm}) \cdot (12.8 \text{ cm}) = 130.56 \text{ cm}^2$$

Then we find the area of the circle. The radius is $\frac{10.2 \text{ cm}}{2}$ = 5.1 cm.

$$\pi \cdot r \cdot r \approx 3.14 \cdot 5.1 \text{ cm} \cdot 5.1 \text{ cm} = 81.6714 \text{ cm}^2$$

Finally we subtract to find the area of the shaded region.

$$130.56 \text{ cm}^2 - 81.6714 \text{ cm}^2 = 48.8886 \text{ cm}^2$$

47. 0.875

a) Move the decimal point   0.87.5
   2 places to the right.

b) Add a percent symbol.   87.5%

   0.875 = 87.5%

49. $0.\overline{6}$

a) Move the decimal point $0.66.\overline{6}$
   2 places to the right.

b) Add a percent symbol. $66.\overline{6}\%$

   $0.\overline{6} = 66.\overline{6}\%$

51. Find 3927 ÷ 1250 using a calculator.

$$\frac{3927}{1250} = 3.1416 \approx 3.142 \qquad \text{Rounding}$$

53. The height of the stack of tennis balls is three times the diameter of one ball, or $3 \cdot d$.

The circumference of one ball is given by $\pi \cdot d$.

The circumference of one ball is greater than the height of the stack of balls, because $\pi > 3$.

Exercise Set 9.7

1. $\sqrt{100} = 10$

The square root of 100 is 10 because 10² = 100.

3. $\sqrt{225} = 15$

The square root of 225 is 15 because 15² = 225.

5. $\sqrt{625} = 25$

The square root of 625 is 25 because 25² = 625.

7. $\sqrt{484} = 22$

The square root of 484 is 22 because 22² = 484.

9. $\sqrt{529} = 23$

The square root of 529 is 23 because 23² = 529.

11. $\sqrt{10,000} = 100$

The square root of 10,000 is 100 because $100^2 = 10,000$.

13. $\sqrt{48} \approx 6.928$

15. $\sqrt{8} \approx 2.828$

17. $\sqrt{18} \approx 4.243$

19. $\sqrt{6} \approx 2.449$

21. $\sqrt{10} \approx 3.162$

23. $\sqrt{75} \approx 8.660$

25. $\sqrt{196} = 14$

27. $\sqrt{183} \approx 13.528$

29.
$$a^2 + b^2 = c^2 \quad \text{Pythagorean equation}$$
$$3^2 + 5^2 = c^2 \quad \text{Substituting}$$
$$9 + 25 = c^2$$
$$34 = c^2$$
$$\sqrt{34} = c \quad \text{Exact answer}$$
$$5.831 \approx c \quad \text{Approximation}$$

31.
$$a^2 + b^2 = c^2 \quad \text{Pythagorean equation}$$
$$7^2 + 7^2 = c^2 \quad \text{Substituting}$$
$$49 + 49 = c^2$$
$$98 = c^2$$
$$\sqrt{98} = c \quad \text{Exact answer}$$
$$9.899 \approx c \quad \text{Approximation}$$

33.
$$a^2 + b^2 = c^2$$
$$a^2 + 12^2 = 13^2$$
$$a^2 + 144 = 169$$
$$a^2 = 169 - 144 = 25$$
$$a = 5$$

35.
$$a^2 + b^2 = c^2$$
$$6^2 + b^2 = 10^2$$
$$36 + b^2 = 100$$
$$b^2 = 100 - 36 = 64$$
$$b = 8$$

37.
$$a^2 + b^2 = c^2$$
$$5^2 + 12^2 = c^2$$
$$25 + 144 = c^2$$
$$169 = c^2$$
$$13 = c$$

39.
$$a^2 + b^2 = c^2$$
$$18^2 + b^2 = 30^2$$
$$324 + b^2 = 900$$
$$b^2 = 900 - 324 = 576$$
$$b = 24$$

41.
$$a^2 + b^2 = c^2$$
$$a^2 + 1^2 = 20^2$$
$$a^2 + 1 = 400$$
$$a^2 = 400 - 1 = 399$$
$$a = \sqrt{399}$$
$$a \approx 19.975$$

43.
$$a^2 + b^2 = c^2$$
$$1^2 + b^2 = 15^2$$
$$1 + b^2 = 225$$
$$b^2 = 225 - 1 = 224$$
$$b = \sqrt{224}$$
$$b \approx 14.967$$

45.
$$a^2 + b^2 = c^2$$
$$12^2 + 5^2 = c^2$$
$$144 + 25 = c^2$$
$$169 = c^2$$
$$13 = c$$

47. <u>Familiarize</u>. We first make a drawing. In it we see a right triangle. We let w = the length of the wire.

<u>Translate</u>. We substitute 9 for a, 13 for b, and w for c in the Pythagorean equation.
$$a^2 + b^2 = c^2$$
$$9^2 + 13^2 = w^2$$

<u>Solve</u>. We solve the equation for w.
$$81 + 169 = w^2$$
$$250 = w^2$$
$$\sqrt{250} = w \quad \text{Exact answer}$$
$$15.811 = w \quad \text{Approximation}$$

<u>Check</u>. $9^2 + 13^2 = 81 + 169 = 250 = (\sqrt{250})^2$

<u>State</u>. The length of the wire is $\sqrt{250}$, or about 15.811 m.

49. Familiarize. We refer to the drawing in the text.
We let d = the distance from home to second base.

Translate. We substitute 65 for a, 65 for b, and
d for c in the Pythagorean equation.

$$a^2 + b^2 = c^2$$
$$65^2 + 65^2 = d^2$$

Solve. We solve the equation for d.

$$4225 + 4225 = d^2$$
$$8450 = d^2$$
$$\sqrt{8450} = d$$
$$91.924 \approx d$$

Check. $65^2 + 65^2 = 4225 + 4225 = 8450 = (\sqrt{8450})^2$

State. The distance from home to second base is
$\sqrt{8450}$, or about 91.924 ft.

51. Familiarize. We refer to the drawing in the text.

Translate. We substitute in the Pythagorean
equation.

$$a^2 + b^2 = c^2$$
$$20^2 + h^2 = 30^2$$

Solve. We solve the equation for h.

$$400 + h^2 = 900$$
$$h^2 = 500$$
$$h = \sqrt{500}$$
$$h \approx 22.361$$

Check. $20^2 + (\sqrt{500})^2 = 400 + 500 = 900 = 30^2$

State. The height of the tree is $\sqrt{500}$, or about
22.361 ft.

53. 45.6%

a) Drop the percent symbol.   45.6

b) Move the decimal point   .45.6
2 places to the left.

45.6% = 0.456

55. 123%

a) Drop the percent symbol   123

b) Move the decimal point   1.23.
2 places to the left.

123% = 1.23

57. To find the areas, we must first use the
Pythagorean equation to find the height of each
triangle and then use the formula for the area
of a triangle.

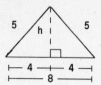

$$a^2 + b^2 = c^2 \qquad \text{Pythagorean equation}$$
$$4^2 + h^2 = 5^2 \qquad \text{Substituting}$$
$$16 + h^2 = 25$$
$$h^2 = 25 - 16 = 9$$
$$h = 3$$

$$A = \frac{1}{2} \cdot b \cdot h \qquad \text{Area of a triangle}$$
$$A = \frac{1}{2} \cdot 8 \cdot 3 \qquad \text{Substituting}$$
$$A = 12$$

$$a^2 + b^2 = c^2 \qquad \text{Pythagorean equation}$$
$$3^2 + h^2 = 5^2 \qquad \text{Substituting}$$
$$9 + h^2 = 25$$
$$h^2 = 25 - 9 = 16$$
$$h = 4$$

$$A = \frac{1}{2} \cdot b \cdot h \qquad \text{Area of a triangle}$$
$$A = \frac{1}{2} \cdot 6 \cdot 4 \qquad \text{Substituting}$$
$$A = 12$$

The areas of the triangles are the same (12 square
units).

Exercise Set 10.1

1. $V = \ell \cdot w \cdot h$
   $V = 12 \text{ cm} \cdot 8 \text{ cm} \cdot 8 \text{ cm}$
   $V = 12 \cdot 64 \text{ cm}^3$
   $V = 768 \text{ cm}^3$

3. $V = \ell \cdot w \cdot h$
   $V = 7.5 \text{ cm} \cdot 2 \text{ cm} \cdot 3 \text{ cm}$
   $V = 7.5 \cdot 6 \text{ cm}^3$
   $V = 45 \text{ cm}^3$

5. $V = \ell \cdot w \cdot h$
   $V = 10 \text{ m} \cdot 5 \text{ m} \cdot 1.5 \text{ m}$
   $V = 10 \cdot 7.5 \text{ m}^3$
   $V = 75 \text{ m}^3$

7. $V = \ell \cdot w \cdot h$
   $V = 6\frac{1}{2} \text{ yd} \cdot 5\frac{1}{2} \text{ yd} \cdot 10 \text{ yd}$
   $V = \frac{13}{2} \cdot \frac{11}{2} \cdot 10 \text{ yd}^3$
   $V = \frac{715}{2} \text{ yd}^3$
   $V = 357\frac{1}{2} \text{ yd}^3$

9. $1 \text{ L} = 1000 \text{ mL} = 1000 \text{ cm}^3$
   These conversion relations appear in the text on page 439.

11. $87 \text{ L} = 87 \times (1 \text{ L})$
    $= 87 \times (1000 \text{ mL})$
    $= 87,000 \text{ mL}$

13. $49 \text{ mL} = 49 \times (1 \text{ mL})$
    $= 49 \times (0.001 \text{ L})$
    $= 0.049 \text{ L}$

15. $0.401 \text{ mL} = 0.401 \times (1 \text{ mL})$
    $= 0.401 \times (0.001 \text{ L})$
    $= 0.000401 \text{ L}$

17. $78.1 \text{ L} = 78.1 \times (1 \text{ L})$
    $= 78.1 \times (1000 \text{ cm}^3)$
    $= 78,100 \text{ cm}^3$

19. $10 \text{ qt} = 10 \times 1 \text{ qt}$
    $= 10 \times 2 \text{ pt}$
    $= 10 \times 2 \times 1 \text{ pt}$
    $= 10 \times 2 \times 16 \text{ oz}$
    $= 320 \text{ oz}$

21. $1 \text{ gal} = 4 \text{ qt}$
    $= 4 \cdot 1 \text{ qt}$
    $= 4 \cdot 2 \text{ pt}$
    $= 4 \cdot 2 \cdot 1 \text{ pt}$
    $= 4 \cdot 2 \cdot 16 \text{ oz}$
    $= 128 \text{ oz}$

23. $8 \text{ gal} = 8 \times 1 \text{ gal}$
    $= 8 \times 4 \text{ qt}$
    $= 32 \text{ qt}$

25. We convert 0.5 L to milliliters:
    $0.5 \text{ L} = 0.5 \times (1 \text{ L})$
    $= 0.5 \times (1000 \text{ mL})$
    $= 500 \text{ mL}$

27. First convert 3 L to milliliters.
    $3 \text{ L} = 3 \times (1 \text{ L})$
    $= 3 \times (1000 \text{ mL})$
    $= 3000 \text{ mL}$
    Then translate to an equation.

    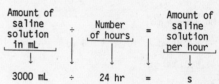

Amount of saline solution in mL	÷	Number of hours	=	Amount of saline solution per hour
↓		↓		↓
3000 mL	÷	24 hr	=	s

    $125 = s$ Carrying out the division

    Thus, 125 mL per hour must be administered.

29. $V = \ell \cdot w \cdot h$
    $V = 18 \text{ yd} \cdot 18 \text{ yd} \cdot 18 \text{ yd}$
    $= 5832 \text{ yd}^3$
    The volume of the world's gold is 5832 yd³.

31. $I = 13\% \times \$600 \times \frac{1}{2}$
    $= 0.13 \times \$600 \times \frac{1}{2}$
    $= \$78 \times \frac{1}{2}$
    $= \$39$

33. $10^3 = 10 \cdot 10 \cdot 10 = 1000$

35. First find the volume of one one-dollar bill in cubic inches:
    $V = \ell \cdot w \cdot h$
    $V = 6.0625 \text{ in.} \times 2.3125 \text{ in.} \times 0.0041 \text{ in.}$
    $V \approx 0.05748 \text{ in}^3$ Rounding

    Then multiply to find the volume of one million one-dollar bills in cubic inches:
    $1,000,000 \times 0.05748 \text{ in}^3 = 57,480 \text{ in}^3$

    The volume of one million one-dollar bills is about 57,480 in³.

35. (continued)

Next we convert to cubic feet:

$57,480 \text{ in}^3 = 57,480 \times \text{in.} \times \text{in.} \times \text{in.}$

$= 57,480 \times \text{in.} \times \text{in.} \times \text{in.} \times$

$\dfrac{1 \text{ ft}}{12 \text{ in.}} \times \dfrac{1 \text{ ft}}{12 \text{ in.}} \times \dfrac{1 \text{ ft}}{12 \text{ in.}}$

$= \dfrac{57,480}{12 \times 12 \times 12} \times \dfrac{\text{in.}}{\text{in.}} \times \dfrac{\text{in.}}{\text{in.}} \times \dfrac{\text{in.}}{\text{in.}} \times$

$\text{ft} \times \text{ft} \times \text{ft}$

$\approx 33.3 \text{ ft}^3$

Exercise Set 10.2

1. $V = Bh = \pi \cdot r^2 \cdot h$

$\approx 3.14 \times 8 \text{ in.} \times 8 \text{ in.} \times 4 \text{ in.}$

$= 803.84 \text{ in}^3$

3. $V = Bh = \pi \cdot r^2 \cdot h$

$\approx 3.14 \times 5 \text{ cm} \times 5 \text{ cm} \times 4.5 \text{ cm}$

$= 353.25 \text{ cm}^3$

5. $V = Bh = \pi \cdot r^2 \cdot h$

$\approx \dfrac{22}{7} \times 210 \text{ yd} \times 210 \text{ yd} \times 300 \text{ yd}$

$= 41,580,000 \text{ yd}^3$

7. $V = \dfrac{4}{3} \cdot \pi \cdot r^3$

$\approx \dfrac{4}{3} \times 3.14 \times (100 \text{ in.})^3$

$= 4,186,666.67 \text{ in}^3$

9. $V = \dfrac{4}{3} \cdot \pi \cdot r^3$

$\approx \dfrac{4}{3} \times 3.14 \times (3.1 \text{ m})^3$

$\approx 124.72 \text{ m}^3$

11. $V = \dfrac{4}{3} \cdot \pi \cdot r^3$

$\approx \dfrac{4}{3} \times \dfrac{22}{7} \times (7 \text{ km})^3$

$\approx \dfrac{4}{3} \times \dfrac{22}{7} \times 343 \text{ km}^3$

$= 1437\dfrac{1}{3} \text{ km}^3$

13. $V = \dfrac{1}{3}\pi \cdot r^2 \cdot h$

$\approx \dfrac{1}{3} \times 3.14 \times 33 \text{ ft} \times 33 \text{ ft} \times 100 \text{ ft}$

$= 113,982 \text{ ft}^3$

15. $V = \dfrac{1}{3}\pi \cdot r^2 \cdot h$

$\approx \dfrac{1}{3} \times \dfrac{22}{7} \times 1.4 \text{ cm} \times 1.4 \text{ cm} \times 12 \text{ cm}$

$= 24.64 \text{ cm}^3$

17. We must find the radius of the base in order to use the formula for the volume of a circular cylinder.

$r = \dfrac{d}{2} = \dfrac{14 \text{ m}}{2} = 7 \text{ m}$

$V = Bh = \pi \cdot r^2 \cdot h$

$\approx \dfrac{22}{7} \times 7 \text{ m} \times 7 \text{ m} \times 220 \text{ m}$

$= 33,880 \text{ m}^3$

19. We must find the radius of the silo in order to use the formula for the volume of a circular cylinder.

$r = \dfrac{d}{2} = \dfrac{6 \text{ m}}{2} = 3 \text{ m}$

$V = Bh = \pi \cdot r^2 \cdot h$

$\approx 3.14 \times 3 \text{ m} \times 3 \text{ m} \times 13 \text{ m}$

$= 367.38 \text{ m}^3$

21. We must find the radius of the tank in order to use the formula for the volume of a sphere.

$r = \dfrac{d}{2} = \dfrac{6 \text{ m}}{2} = 3 \text{ m}$

$V = \dfrac{4}{3} \cdot \pi \cdot r^3$

$\approx \dfrac{4}{3} \times 3.14 \times (3 \text{ m})^3$

$\approx 113.0 \text{ m}^3$ Rounding to the nearest tenth of a cubic meter.

23. We must find the radius of the earth in order to use the formula for the volume of a sphere.

$r = \dfrac{d}{2} = \dfrac{6400 \text{ km}}{2} = 3200 \text{ km}$

$V = \dfrac{4}{3} \cdot \pi \cdot r^3$

$\approx \dfrac{4}{3} \cdot 3.14 \times (3200 \text{ km})^3$

$\approx 137,188,693,333.33 \text{ km}^3$

25. First find the radius of the tank in inches:

$r = \dfrac{d}{2} = \dfrac{16 \text{ in.}}{2} = 8 \text{ in.}$

Then convert 8 in. to feet:

$8 \text{ in.} = 8 \text{ in.} \times \dfrac{1 \text{ ft}}{12 \text{ in.}} = \dfrac{8}{12} \times \dfrac{\text{in.}}{\text{in.}} \times \text{ft} = \dfrac{2}{3} \text{ ft}$

Find the volume of the tank:

$V = Bh = \pi \cdot r^2 \cdot h$

$V \approx 3.14 \times \dfrac{2}{3} \text{ ft} \times \dfrac{2}{3} \text{ ft} \times 5 \text{ ft} = 6.9\overline{7} \text{ ft}^3$

The volume of the tank is $6.9\overline{7} \text{ ft}^3$.

We multiply to find the number of gallons the tank will hold:

$6.9\overline{7} \times 7.5 \text{ gal} = 52.\overline{3} \text{ gal}$

The tank will hold $52.\overline{3}$ gallons of water.

Exercise Set 10.3

<u>1.</u> 1 T = 2000 lb

This conversion relation is given in the text on page 447.

<u>3.</u> 6000 lb = 6000 lb × $\frac{1\ T}{2000\ lb}$ Multiplying by 1 using $\frac{1\ T}{2000\ lb}$

= $\frac{6000}{2000}$ × $\frac{lb}{lb}$ × 1 T

= 3 × 1 T   The $\frac{lb}{lb}$ acts like 1, so we can omit it.

= 3 T

<u>5.</u> 4 lb = 4 × 1 lb

= 4 × 16 oz   Substituting 16 oz for 1 lb

= 64 oz

<u>7.</u> 3.5 T = 3.5 × 1 T

= 3.5 × 2000 lb   Substituting 2000 lb for 1 T

= 7000 lb

<u>9.</u> 3200 oz = 3200 oz × $\frac{1\ lb}{16\ oz}$ Multiplying by 1 using $\frac{1\ lb}{16\ oz}$

= $\frac{3200}{16}$ × $\frac{oz}{oz}$ × 1 lb

= 200 × 1 lb   The $\frac{oz}{oz}$ acts like 1, so we can omit it.

= 200 lb

= 200 lb × $\frac{1\ T}{2000\ lb}$ Multiplying by 1 using $\frac{1\ T}{2000\ lb}$

= $\frac{200}{2000}$ × $\frac{lb}{lb}$ × 1 T

= 0.1 × 1 T   The $\frac{lb}{lb}$ acts like 1, so we can omit it.

= 0.1 T

<u>11.</u> 96 oz = 96 oz × $\frac{1\ lb}{16\ oz}$ Multiplying by 1 using $\frac{1\ lb}{16\ oz}$

= $\frac{96}{16}$ × $\frac{oz}{oz}$ × 1 lb

= 6 × 1 lb

= 6 lb

<u>13.</u> 1 kg = _____ g

Think: To go from kg to g in the table is a move of 3 places to the right. Thus, we move the decimal point 3 places to the right.

1   1.000.

1 kg = 1000 g

<u>15.</u> 1 dag = _____ g

Think: To go from dag to g in the table is a move of 1 place to the right. Thus, we move the decimal point 1 place to the right.

1   1.0.

1 dag = 10 g

<u>17.</u> 1 cg = _____ g

Think: To go from cg to g in the table is a move of 2 places to the left. Thus, we move the decimal point 2 places to the left.

1   .01.

1 cg = 0.01 g

<u>19.</u> 1 g = _____ mg

Think: To go from g to mg in the table is a move of 3 places to the right. Thus, we move the decimal point 3 places to the right.

1   1.000.

1 g = 1000 mg

<u>21.</u> 1 g = _____ dg

Think: To go from g to dg in the table is a move of 1 place to the right. Thus, we move the decimal point 1 place to the right.

1   1.0.

1 g = 10 dg

<u>23.</u> Complete: 234 kg = _____ g

Think: To go from kg to g in the table is a move of 3 places to the right. Thus, we move the decimal point 3 places to the right.

234   234.000.

234 kg = 234,000 g

<u>25.</u> Complete: 5200 g = _____ kg

Think: To go from g to kg in the table is a move of 3 places to the left. Thus, we move the decimal point 3 places to the left.

5200   5.200.

5200 g = 5.2 kg

27. Complete: 67 hg = _____ kg

    Think: To go from hg to kg in the table is a move of 1 place to the left. Thus, we move the decimal point 1 place to the left.

        67   6.7.

        67 hg = 6.7 kg

29. Complete: 0.502 dg = _____ g

    Think: To go from dg to g in the table is a move of 1 place to the left. Thus, we move the decimal point 1 place to the left.

        0.502   0.0.502

        0.502 dg = 0.0502 g

31. 6780 g = _____ kg

    Think: To go from g to kg in the table is a move of 3 places to the left. Thus, we move the decimal point 3 places to the left.

        6780   6.780.

        6780 g = 6.78 kg

33. Complete: 69 mg = _____ cg

    Think: To go from mg to cg in the table is a move of 1 place to the left. Thus, we move the decimal point 1 place to the left.

        69   6.9.

        69 mg = 6.9 cg

35. Complete: 8 kg = _____ cg

    Think: To go from kg to cg in the table is a move of 5 places to the right. Thus, we move the decimal point 5 places to the right.

        8   8.00000.

        8 kg = 800,000 cg

37. 1 t = 1000 kg

    This conversion relation is given in the text on page 448.

39. Complete: 3.4 cg = _____ dag

    Think: To go from cg to dag in the table is a move of 3 places to the left. Thus, we move the decimal point 3 places to the left.

        3.4   0.003.4

        3.4 cg = 0.0034 dag

41. 1 day = 24 hr

    This conversion relation is given in the text on page 450.

43. 1 min = 60 sec

    This conversion relation is given in the text on page 450.

45. 1 yr = 365 $\frac{1}{4}$ days

    This conversion relation is given in the text on page 450.

47. 2 wk = 2 × 1 wk

    = 2 × 7 days   Substituting 7 days for 1 wk

    = 14 days

    = 14 × 1 day

    = 14 × 24 hr   Substituting 24 hr for 1 day

    = 336 hr

49. 492 sec = 492 sec × $\frac{1 \text{ min}}{60 \text{ sec}}$ Multiplying by 1 using $\frac{1 \text{ min}}{60 \text{ sec}}$

    = $\frac{492}{60}$ × $\frac{\text{sec}}{\text{sec}}$ × 1 min

    = 8.2 × 1 min   The $\frac{\text{sec}}{\text{sec}}$ acts like 1, so we can omit it.

    = 8.2 min

51. $2^4 = 2 \cdot 2 \cdot 2 \cdot 2 = 16$

53. $5^3 = 5 \cdot 5 \cdot 5 = 125$

55. 1 lb = 453.5 g   Using the conversion given
    453.5 g = _____ kg

    Think: To go from g to kg in the table is a move of 3 places to the left. Thus, we move the decimal point 3 places to the left.

        453.5   .453.5

        1 lb = 0.4535 kg

57. 1 mg = $\frac{1}{1000}$ g   Using the conversion on page 448 of the text

    $\frac{1}{1000}$ g = 1000 × $\frac{1}{1,000,000}$ g

    = 1000 × μg   Using the definition of μg

    = 1000 μg

    1 mg = 1000 μg

59. Convert 125 µg to milligrams.

$$125 \text{ µg} = 125 \times 1 \text{ µg}$$

$$= 125 \times \frac{1}{1,000,000} \text{ g}$$

$$= 125 \times \frac{1}{1000} \times \frac{1}{1000} \text{ g}$$

$$= \frac{125}{1000} \times 1 \text{ mg}$$

$$= 0.125 \text{ mg, or } \frac{1}{8} \text{ mg}$$

125 µg = 0.125 mg

61. First convert 500 mg to grams.

Think: To go from mg to g in the table is a move of 3 places to the left. Thus, we move the decimal point 3 places to the left.

500   .500.

500 mg = 0.5 g

Now translate to an equation.

$$\begin{pmatrix} \text{Dosage per} \\ \text{tablet} \\ \text{in g} \end{pmatrix} \text{ times } \begin{pmatrix} \text{Number of} \\ \text{tablets} \end{pmatrix} \text{ is } \begin{pmatrix} \text{Total} \\ \text{dosage} \\ \text{in g} \end{pmatrix}$$

$$0.5 \qquad \times \qquad n \qquad = \qquad 2$$

To solve we divide on both sides by 0.5.

$$n = 2 \div 0.5$$

$$n = 4$$

Thus, 4 tablets would have to be taken.

63. We solve using a proportion.

$$\text{cephalexin} \longrightarrow \frac{250}{5} = \frac{400}{a} \longleftarrow \text{cephalexin}$$
$$\text{liquid} \longrightarrow \qquad\qquad\quad \longleftarrow \text{liquid}$$

Solve: $250 \cdot a = 5 \cdot 400$

$$a = \frac{5 \cdot 400}{250}$$

$$a = 8$$

Thus, 8 mL of liquid would be required.

65. First, use a proportion to determine how many eggs are in one pound.

$$\text{eggs} \longrightarrow \frac{12}{0.75} = \frac{x}{1.50} \longleftarrow \text{eggs}$$
$$\text{price} \longrightarrow \qquad\qquad\qquad \longleftarrow \text{price}$$

$$12 \cdot 1.50 = 0.75 \cdot x$$

$$\frac{12 \cdot 1.50}{0.75} = x$$

$$24 = x$$

There are 24 eggs in one pound.

65. (continued)

Then use a proportion to find the weight of one egg.

$$\text{weight} \longrightarrow \frac{16}{24} = \frac{w}{1} \longleftarrow \text{weight}$$
$$\text{eggs} \longrightarrow \qquad\qquad\quad \longleftarrow \text{eggs}$$

$$16 \cdot 1 = 24 \cdot w$$

$$\frac{16 \cdot 1}{24} = w$$

$$\frac{2}{3} = w$$

One egg weighs $\frac{2}{3}$ oz.

67. Use a calculator.

1,000,000 sec

$$= 1,000,000,000 \text{ sec} \times \frac{1 \text{ min}}{60 \text{ sec}} \times \frac{1 \text{ hr}}{60 \text{ min}} \times \frac{1 \text{ day}}{24 \text{ hr}} \times$$

$$\frac{1 \text{ yr}}{365\frac{1}{4} \text{ day}}$$

≈ 31.7 Rounded to the nearest tenth

One billion seconds is approximately 31.7 years.

Exercise Set 10.4

1. By laying a ruler or piece of paper horizontally between the scales on page 455, we see that 178° F ≈ 80° C.

3. By laying a ruler or piece of paper horizontally between the scales on page 455, we see that 140° F ≈ 60° C.

5. By laying a ruler or piece of paper horizontally between the scales on page 455, we see that 88° F ≈ 30° C.

7. By laying a ruler or piece of paper horizontally between the scales on page 455, we see that 10° F ≈ -10° C.

9. By laying a ruler or piece of paper horizontally between the scales on page 455, we see that 86° C ≈ 190° F.

11. By laying a ruler or piece of paper horizontally between the scales on page 455, we see that 58° C ≈ 140° F.

13. By laying a ruler or piece of paper horizontally between the scales on page 455, we see that -10° C ≈ 10° F.

15. By laying a ruler or piece of paper horizontally between the scales on page 455, we see that 5° C ≈ 40° F.

17. $F = \frac{9}{5} \cdot C + 32$

$$F = \frac{9}{5} \cdot 25 + 32 = 45 + 32 = 77$$

Thus, 25° C = 77° F.

19. $F = \frac{9}{5} \cdot C + 32$

  $F = \frac{9}{5} \cdot 40 + 32 = 72 + 32 = 104$

  Thus, 40° C = 104° F.

21. $F = \frac{9}{5} \cdot C + 32$

  $F = \frac{9}{5} \cdot 3000 + 32 = 5400 + 32 = 5432$

  Thus, 3000° C = 5432° F.

23. $C = \frac{5}{9} \cdot (F - 32)$

  $C = \frac{5}{9} \cdot (86 - 32) = \frac{5}{9} \cdot 54 = 30$

  Thus, 86° F = 30° C.

25. $C = \frac{5}{9} \cdot (F - 32)$

  $C = \frac{5}{9} \cdot (131 - 32) = \frac{5}{9} \cdot 99 = 55$

  Thus, 131° F = 55° C.

27. $C = \frac{5}{9} \cdot (F - 32)$

  $C = \frac{5}{9} (98.6 - 32) = \frac{5}{9} \cdot 66.6 = 37$

  Thus, 98.6° F = 37° C.

29. 23.4 cm = _____ mm

  Think: To go from cm to mm in the table is a move of 1 place to the right. Thus, we move the decimal point 1 place to the right.

  23.4   23.4.
                ⌞↑

  23.4 cm = 234 mm

31. 28 ft = 28 × 1 ft = 28 × 12 in. = 336 in.

33. We first convert from Kelvin to Celsius temperature and then from Celsius to Fahrenheit temperature.

  K = C + 273
  400 = C + 273   Substituting
  127 = C   Subtracting 273 on both sides
  Thus, 400° Kelvin = 127° C.

  $F = \frac{9}{5} \cdot C + 32$

  $F = \frac{9}{5} \cdot 127 + 32 = 228.6 + 32 = 260.6$

  Thus, 127° C = 260.6° F, so the reaction will take place at 260.6° F.

Exercise Set 10.5

1. 1 ft² = 144 in²

  This conversion relation is given in the text on page 459.

3. 1 mi² = 640 acres

  This conversion relation is given in the text on page 459.

5. $1 \text{ in}^2 = 1 \text{ in}^2 \times \frac{1 \text{ ft}^2}{144 \text{ in}^2}$ Multiplying by 1 using $\frac{1 \text{ ft}^2}{144 \text{ in}^2}$

  $= \frac{1}{144} \times \frac{\text{in}^2}{\text{in}^2} \times 1 \text{ ft}^2$

  $= \frac{1}{144} \text{ ft}^2$

7. 22 yd² = 22 × 1 yd²
    = 22 × 9 ft²   Substituting 9 ft² for 1 yd²
    = 198 ft²

9. 44 yd² = 44·1 yd²
    = 44·9 ft²   Substituting 9 ft² for 1 yd²
    = 396 ft²

11. 20 mi² = 20·1 mi²
    = 20·640 acres   Substituting 640 acres for 1 mi²
    = 12,800 acres

13. 1 mi² = 1·(1 mi)²
    = 1·(5280 ft)²   Substituting 5280 ft for 1 mi
    = 5280 ft·5280 ft
    = 27,878,400 ft²

15. $720 \text{ in}^2 = 720 \text{ in}^2 \cdot \frac{1 \text{ ft}^2}{144 \text{ in}^2}$ Multiplying by 1 using $\frac{1 \text{ ft}^2}{144 \text{ in}^2}$

  $= \frac{720}{144} \times \frac{\text{in}^2}{\text{in}^2} \times 1 \text{ ft}^2$

  $= 5 \text{ ft}^2$

17. 144 in² = 1 ft²

  This conversion relation is given in the text on page 459.

19. First we find the area of the rectangle:
    A = ℓ·w = 12 ft·18 ft = 216 ft²
  We find the area of one square:
    A = s·s = 3 in. · 3 in. = 9 in²
  Then we multiply to find the area of 20 squares:
    20·9 in² = 180 in²

19. (continued)

We convert the area of the square to square feet:

$$180 \text{ in}^2 = 180 \text{ in}^2 \cdot \frac{1 \text{ ft}^2}{144 \text{ in}^2} = \frac{180}{144} \cdot \frac{\text{in}^2}{\text{in}^2} \cdot$$

$$1 \text{ ft}^2 = 1.25 \text{ ft}^2$$

Finally we subtract to find the area of the sheet, in square feet:

$$216 \text{ ft}^2 - 1.25 \text{ ft}^2 = 214.75 \text{ ft}^2$$

21. 20 km² = _____ m²

Think: To go from km to m in the table is a move of 3 places to the right. So we move the decimal point 2·3, or 6 places to the right.

20   20.000000.

20 km² = 20,000,000 m²

23. 0.014 m² = _____ cm²

Think: To go from m to cm in the table is a move of 2 places to the right. So we move the decimal point 2·2, or 4 places to the right.

0.014   0.0140.

0.014 m² = 140 cm²

25. 2345.6 mm² = _____ cm²

Think: To go from mm to cm in the table is a move of 1 place to the left. So we move the decimal point 2·1, or 2 places to the left.

2345.6   23.45.6

2345.6 mm² = 23.456 cm²

27. 1200 cm² = _____ m²

Think: To go from cm to m in the table is a move of 2 places to the left. Thus, we move the decimal point 2·2, or 4 places to the left.

1200   .1200.

1200 cm² = 0.12 m² .

29. 250,000 mm² = _____ cm²

Think: To go from mm to cm in the table is a move of 1 place to the left. Thus, we move the decimal point 2·1, or 2 places to the left.

250,000   2500.00.

250,000 mm² = 2500 cm²

31. We convert the base and height to feet and then find the area.

5 yd = 5 × 1 yd = 5 × 3 ft = 15 ft

$$3 \text{ in.} = 3 \text{ in.} \times \frac{1 \text{ ft}}{12 \text{ in.}} = \frac{3}{12} \times \frac{\text{in.}}{\text{in.}} \times 1 \text{ ft} =$$

0.25 ft

$$A = \frac{1}{2} \cdot b \cdot h$$

$$A = \frac{1}{2} \cdot 15 \text{ ft} \cdot 0.25 \text{ ft} = 1.875 \text{ ft}^2$$

The area of the shaded region is 1.875 ft².

33. We convert the height to feet and then find the area.

$$10 \text{ in.} = 10 \text{ in.} \times \frac{1 \text{ ft}}{12 \text{ in.}} = \frac{10}{12} \times \frac{\text{in.}}{\text{in.}} \times 1 \text{ ft} = \frac{5}{6} \text{ ft}$$

$$A = b \cdot h$$

$$A = 16 \text{ ft} \cdot \frac{5}{6} \text{ ft} = \frac{16 \cdot 5}{6} \text{ ft}^2 = \frac{40}{3} \text{ ft}^2 = 13\frac{1}{3} \text{ ft}^2$$

The area of the shaded region is $13\frac{1}{3}$ ft².

35. The shaded area consists of one triangle, with base 8.4 cm and height 10 cm, and seven triangles, each with base and height 1.3 mm. We will express the area in square centimeters.

We find the area of the large triangle:

$$A = \frac{1}{2} \cdot b \cdot h$$

$$A = \frac{1}{2} \cdot 8.4 \text{ cm} \cdot 10 \text{ cm} = 42 \text{ cm}^2$$

We convert the dimensions of the small triangles to centimeters. To go from mm to cm in the table is a move of 1 place to the left. Thus, we move the decimal point one place to the left.

1.3   .1.3

1.3 mm = 0.13 cm

Then we find the area of one of the small triangles:

$$A = \frac{1}{2} \cdot b \cdot h$$

$$A = \frac{1}{2} \cdot 0.13 \text{ cm} \cdot 0.13 \text{ cm} = 0.00845 \text{ cm}^2$$

We multiply to find the area of the seven small triangles:

$$7 \cdot 0.00845 \text{ cm}^2 = 0.05915 \text{ cm}^2$$

Finally we add to find the area of the shaded region:

$$42 \text{ cm}^2 + 0.05915 \text{ cm}^2 = 42.05915 \text{ cm}^2$$

The area of the shaded region is 42.05915 cm².

37. $22{,}176 \text{ ft} = 22{,}176 \text{ ft} \times \frac{1 \text{ mi}}{5280 \text{ ft}}$ Multiplying by 1 using $\frac{1 \text{ mi}}{5280 \text{ ft}}$

$$= \frac{22{,}176}{5280} \times \frac{\text{ft}}{\text{ft}} \times 1 \text{ mi}$$

$$= 4.2 \text{ mi}$$

<u>39</u>. Use a calculator.

$$1 \text{ acre} = 1 \text{ acre} \times \frac{1 \text{ mi}^2}{640 \text{ acres}}$$

$$= \frac{1}{640} \times \frac{\text{acre}}{\text{acre}} \times 1 \text{ mi}^2$$

$$= \frac{1}{640} \text{ mi}^2, \text{ or } 0.0015625 \text{ mi}^2$$

Exercise Set 11.1

1. The integer 5 corresponds to winning 5 points, and the integer -12 corresponds to losing 12 points.

3. The integer -1286 corresponds to 1286 ft below sea level. The integer 29,028 corresponds to 29,028 ft above sea level.

5. The integer 750 corresponds to a $750 deposit, and the integer -125 corresponds to a $125 withdrawal.

7. The integers 20, -150, and 300 correspond to the interception of the missile, the loss of the starship, and the capture of the base, respectively.

9. The number $\frac{10}{3}$ can be named $3\frac{1}{3}$, or $3.3\overline{3}$. The graph is $\frac{1}{3}$ of the way from 3 to 4.

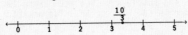

11. The graph of -4.3 is $\frac{3}{10}$ of the way from -4 to -5.

13. See Example 7 in the text.

15. We first find decimal notation for $\frac{5}{3}$. Since $\frac{5}{3}$ means $5 \div 3$, we divide.

$$\begin{array}{r} 1.6\ 6.\,.\,. \\ 3\overline{)5.0\ 0} \\ \underline{3} \\ 2\ 0 \\ \underline{1\ 8} \\ 2\ 0 \\ \underline{1\ 8} \\ 2 \end{array}$$

Thus, $\frac{5}{3} = 1.\overline{6}$, so $-\frac{5}{3} = -1.\overline{6}$.

17. We first find decimal notation for $\frac{7}{6}$. Since $\frac{7}{6}$ means $7 \div 6$, we divide.

$$\begin{array}{r} 1.1\ 6\ 6.\,.\,. \\ 6\overline{)7.0\ 0\ 0} \\ \underline{6} \\ 1\ 0 \\ \underline{6} \\ 4\ 0 \\ \underline{3\ 6} \\ 4\ 0 \\ \underline{3\ 6} \\ 4 \end{array}$$

Thus, $\frac{7}{6} = 1.1\overline{6}$, so $-\frac{7}{6} = -1.1\overline{6}$.

19. We first find decimal notation for $\frac{7}{8}$. Since $\frac{7}{8}$ means $7 \div 8$, we divide.

$$\begin{array}{r} 0.8\ 7\ 5 \\ 8\overline{)7.0\ 0\ 0} \\ \underline{6\ 4} \\ 6\ 0 \\ \underline{5\ 6} \\ 4\ 0 \\ \underline{4\ 0} \\ 0 \end{array}$$

Thus, $\frac{7}{8} = 0.875$, so $-\frac{7}{8} = -0.875$.

21. We first find decimal notation for $\frac{7}{20}$. Since $\frac{7}{20}$ means $7 \div 20$, we divide.

$$\begin{array}{r} 0.3\ 5 \\ 2\ 0\overline{)7.0\ 0} \\ \underline{6\ 0} \\ 1\ 0\ 0 \\ \underline{1\ 0\ 0} \\ 0 \end{array}$$

Thus, $\frac{7}{20} = 0.35$, so $-\frac{7}{20} = -0.35$.

23. Since 9 is to the right of 0, we have 9 > 0.

25. Since 8 is to the right of -8, we have 8 > -8.

27. Since 0 is to the right of -7, we have 0 > -7.

29. Since -4 is to the left of -3, we have -4 < -3.

31. Since -3 is to the right of -4, we have -3 > -4.

33. Since -10 is to the right of -14, we have -10 > -14.

35. Since -3.3 is to the left of -2.2, we have -3.3 < -2.2.

37. Since 17.2 is to the right of -1.67, we have 17.2 > -1.67.

39. Since -14.34 is to the right of -17.88, we have -14.34 > -17.88.

41. We convert to decimal notation: $-\frac{14}{17} \approx -0.8235$, and $-\frac{27}{35} \approx -0.7714$. Since -0.8235 is to the left of -0.7714, we have $-\frac{14}{17} < -\frac{27}{35}$.

43. The distance of -7 from 0 is 7, so $|-7| = 7$.

45. The distance of 11 from 0 is 11, so $|11| = 11$.

47. The distance of -4 from 0 is 4, so $|-4| = 4$.

49. The distance of 325 from 0 is 325, so $|325| = 325$.

51. The distance of $-\frac{10}{7}$ from 0 is $\frac{10}{7}$, so $\left|-\frac{10}{7}\right| = \frac{10}{7}$.

53. The distance of 14.8 from 0 is 14.8, so
|14.8| = 14.8.

55.
```
          3
       2⟌6
     2⟌1 2
    2⟌2 4
   2⟌4 8
  2⟌9 6
 2⟌1 9 2
```

192 = 2·2·2·2·2·2·3

57. 6 = 2·3

24 = 2·2·2·3

32 = 2·2·2·2·2

The LCM is 2·2·2·2·2·3, or 96.

59. |4| = 4, and |-7| = 7. Since 4 is to the left of 7, we have |4| < |-7|.

61. $-\frac{2}{3}, \frac{1}{2}, -\frac{3}{4}, -\frac{5}{6}, \frac{3}{8}, \frac{1}{6}$ can be written in decimal notation as $-0.\overline{6}, 0.5, -0.75, -0.8\overline{3}, 0.375, 0.1\overline{6}$, respectively. Listing from least to greatest, we have $-\frac{5}{6}, -\frac{3}{4}, -\frac{2}{3}, \frac{1}{6}, \frac{3}{8}, \frac{1}{2}$.

Exercise Set 11.2

1. -9 + 2   The absolute values are 9 and 2. The difference is 9 - 2, or 7. The negative number has the larger absolute value, so the answer is negative. -9 + 2 = -7

3. -10 + 6   The absolute values are 10 and 6. The difference is 10 - 6, or 4. The negative number has the larger absolute value, so the answer is negative. -10 + 6 = -4

5. -8 + 8   A positive and a negative number. The numbers have the same absolute value. The sum is 0. -8 + 8 = 0

7. -3 + (-5)   Two negatives. Add the absolute values, getting 8. Make the answer negative. -3 + (-5) = -8

9. -7 + 0   One number is 0. The answer is the other number. -7 + 0 = -7

11. 0 + (-27)   One number is 0. The answer is the other number. 0 + (-27) = -27

13. 17 + (-17)   A positive and a negative number. The numbers have the same absolute value. The sum is 0. 17 + (-17) = 0

15. -17 + (-25)   Two negatives. Add the absolute values, getting 42. Make the answer negative. -17 + (-25) = -42

17. 18 + (-18)   A positive and a negative number. The numbers have the same absolute value. The sum is 0. 18 + (-18) = 0

19. -18 + 18   A positive and a negative number. The numbers have the same absolute value. The sum is 0.   -18 + 18 = 0

21. 8 + (-5)   The absolute values are 8 and 5. The difference is 8 - 5, or 3. The positive number has the larger absolute value, so the answer is positive. 8 + (-5) = 3

23. -4 + (-5)   Two negatives. Add the absolute values, getting 9. Make the answer negative. -4 + (-5) = -9

25. 13 + (-6)   The absolute values are 13 and 6. The difference is 13 - 6, or 7. The positive number has the larger absolute value, so the answer is positive. 13 + (-6) = 7

27. -25 + 25   A positive and a negative number. The numbers have the same absolute value. The sum is 0.   -25 + 25 = 0

29. 63 + (-18)   The absolute values are 63 and 18. The difference is 63 - 18, or 45. The positive number has the larger absolute value, so the answer is positive. 63 + (-18) = 45

31. -6.5 + 4.7   The absolute values are 6.5 and 4.7. The difference is 6.5 - 4.7, or 1.8. The negative number has the larger absolute value, so the answer is negative. -6.5 + 4.7 = -1.8

33. -2.8 + (-5.3)   Two negatives. Add the absolute values, getting 8.1. Make the answer negative. -2.8 + (-5.3) = -8.1

35. $-\frac{3}{5} + \frac{2}{5}$ The absolute values are $\frac{3}{5}$ and $\frac{2}{5}$. The difference is $\frac{3}{5} - \frac{2}{5}$, or $\frac{1}{5}$. The negative number has the larger absolute value, so the answer is negative. $-\frac{3}{5} + \frac{2}{5} = -\frac{1}{5}$

37. $-\frac{3}{7} + \left(-\frac{5}{7}\right)$ Two negatives. Add the absolute values, getting $\frac{8}{7}$. Make the answer negative. $-\frac{3}{7} + \left(-\frac{5}{7}\right) = -\frac{8}{7}$

39. $-\frac{5}{8} + \frac{1}{4}$ The absolute values are $\frac{5}{8}$ and $\frac{1}{4}$. The difference is $\frac{5}{8} - \frac{2}{8}$, or $\frac{3}{8}$. The negative number has the larger absolute value, so the answer is negative. $-\frac{5}{8} + \frac{1}{4} = -\frac{3}{8}$

41. $-\frac{3}{7} + \left(-\frac{2}{5}\right)$ Two negatives. Add the absolute values, getting $\frac{15}{35} + \frac{14}{35}$, or $\frac{29}{35}$. Make the answer negative. $-\frac{3}{7} + \left(-\frac{2}{5}\right) = -\frac{29}{35}$

43. $-\frac{3}{5} + \left(-\frac{2}{15}\right)$ Two negatives. Add the absolute values, getting $\frac{9}{15} + \frac{2}{15}$, or $\frac{11}{15}$. Make the answer negative. $-\frac{3}{5} + \left(-\frac{2}{15}\right) = -\frac{11}{15}$

<u>45.</u> -5.7 + (-7.2) + 6.6 = -12.9 + 6.6   Adding the
                                              positive
                                              numbers

                              = -6.3   Adding the results

<u>47.</u> $-\frac{7}{16} + \frac{7}{8}$ The absolute values are $\frac{7}{16}$ and $\frac{7}{8}$.
The difference is $\frac{14}{16} - \frac{7}{16}$, or $\frac{7}{16}$. The positive
number has the larger absolute value, so the
answer is positive. $-\frac{7}{16} + \frac{7}{8} = \frac{7}{16}$

<u>49.</u> 75 + (-14) + (-17) + (-5)

   a) -14 + (-17) + (-5) = -36   Adding the
                                   negative numbers

   b) 75 + (-36) = 39   Adding the results

<u>51.</u> $-44 + \left(-\frac{3}{8}\right) + 95 + \left(-\frac{5}{8}\right)$

   a) $-44 + \left(-\frac{3}{8}\right) + \left(-\frac{5}{8}\right) = -45$ Adding the
                                                                            negative numbers

   b) -45 + 95 = 50   Adding the results

<u>53.</u> 98 + (-54) + 113 + (-998) + 44 + (-612) + (-18) +
334

   a) 98 + 113 + 44 + 334 = 589   Adding the
                                    positive numbers

   b) -54 + (-998) + (-612) + (-18) = -1682

                             Adding the negative numbers

   c) 589 + (-1682) = -1093   Adding the results

<u>55.</u> The additive inverse of 24 is -24 because 24 +
(-24) = 0.

<u>57.</u> The additive inverse of -26.9 is 26.9 because
-26.9 + 26.9 = 0.

<u>59.</u> If x = 9, then -x = -(9) = -9.   (The additive
                                         inverse of 9 is
                                         -9.)

<u>61.</u> If $x = -\frac{14}{3}$, then $-x = -\left(-\frac{14}{3}\right) = \frac{14}{3}$.
        $\left[\text{The additive inverse of } -\frac{14}{3} \text{ is } \frac{14}{3}.\right]$

<u>63.</u> If x = -65, then -(-x) = -[-(-65)] = -65
        (The opposite of the opposite of -65 is -65.)

<u>65.</u> If $x = \frac{5}{3}$, then $-(-x) = -\left(-\frac{5}{3}\right) = \frac{5}{3}$.
        $\left[\text{The opposite of the opposite of } \frac{5}{3} \text{ is } \frac{5}{3}.\right]$

<u>67.</u> -(-14) = 14

<u>69.</u> -(10) = -10

<u>71.</u> When x is positive, the inverse of x, -x, is
negative.

<u>73.</u> Positive (The sum of two positive numbers is
positive.)

<u>75.</u> If n is positive, -n is negative. Thus -n + m,
the sum of two negatives, is negative.

Exercise Set 11.3

<u>1.</u> 3 - 7 = 3 + (-7) = -4

<u>3.</u> 0 - 7 = 0 + (-7) = -7

<u>5.</u> -8 - (-2) = -8 + 2 = -6

<u>7.</u> -10 - (-10) = -10 + 10 = 0

<u>9.</u> 12 - 16 = 12 + (-16) = -4

<u>11.</u> 20 - 27 = 20 + (-27) = -7

<u>13.</u> -9 - (-3) = -9 + 3 = -6

<u>15.</u> -11 - (-11) = -11 + 11 = 0

<u>17.</u> 8 - (-3) = 8 + 3 = 11

<u>19.</u> -6 - 8 = -6 + (-8) = -14

<u>21.</u> -4 - (-9) = -4 + 9 = 5

<u>23.</u> 2 - 9 = 2 + (-9) = -7

<u>25.</u> 0 - 5 = 0 + (-5) = -5

<u>27.</u> -5 - (-2) = -5 + 2 = -3

<u>29.</u> 2 - 25 = 2 + (-25) = -23

<u>31.</u> -42 - 26 = -42 + (-26) = -68

<u>33.</u> -71 - 2 = -71 + (-2) = -73

<u>35.</u> 24 - (-92) = 24 + 92 = 116

<u>37.</u> -2.8 - 0 = -2.8 + 0 = -2.8

<u>39.</u> $\frac{3}{8} - \frac{5}{8} = \frac{3}{8} + \left(-\frac{5}{8}\right) = -\frac{2}{8} = -\frac{1}{4}$

<u>41.</u> $\frac{3}{4} - \frac{2}{3} = \frac{9}{12} - \frac{8}{12} = \frac{9}{12} + \left(-\frac{8}{12}\right) = \frac{1}{12}$

<u>43.</u> $-\frac{3}{4} - \frac{2}{3} = -\frac{9}{12} - \frac{8}{12} = -\frac{9}{12} + \left(-\frac{8}{12}\right) = -\frac{17}{12}$

<u>45.</u> $-\frac{5}{8} - \left(-\frac{3}{4}\right) = -\frac{5}{8} - \left(-\frac{6}{8}\right) = -\frac{5}{8} + \frac{6}{8} = \frac{1}{8}$

<u>47.</u> 6.1 - (-13.8) = 6.1 + 13.8 = 19.9

<u>49.</u> -3.2 - 5.8 = -3.2 + (-5.8) = -9

51. $0.99 - 1 = 0.99 + (-1) = -0.01$

53. $3 - 5.7 = 3 + (-5.7) = -2.7$

55. $7 - 10.53 = 7 + (-10.53) = -3.53$

57. $\frac{1}{6} - \frac{2}{3} = \frac{1}{6} - \frac{4}{6} = \frac{1}{6} + \left(-\frac{4}{6}\right) = -\frac{3}{6} = -\frac{1}{2}$

59. $-\frac{4}{7} - \left(-\frac{10}{7}\right) = -\frac{4}{7} + \frac{10}{7} = \frac{6}{7}$

61. $-\frac{7}{10} - \frac{10}{15} = -\frac{21}{30} - \frac{20}{30} = -\frac{21}{30} + \left(-\frac{20}{30}\right) = -\frac{41}{30}$

63. $\frac{1}{13} - \frac{1}{12} = \frac{12}{156} - \frac{13}{156} = \frac{12}{156} + \left(-\frac{13}{156}\right) = -\frac{1}{156}$

65. $18 - (-15) - 3 - (-5) + 2 =$
    $18 + 15 + (-3) + 5 + 2 = 37$

67. $-31 + (-28) - (-14) - 17 =$
    $(-31) + (-28) + 14 + (-17) = -62$

69. $-93 - (-84) - 41 - (-56) =$
    $(-93) + 84 + (-41) + 56 = 6$

71. $-5 - (-30) + 30 + 40 - (-12) =$
    $(-5) + 30 + 30 + 40 + 12 = 107$

73. $132 - (-21) + 45 - (-21) = 132 + 21 + 45 + 21 = 219$

75. $A = \ell \cdot w$
    $A = (8.4 \text{ cm}) \times (11.5 \text{ cm})$
    $A = (8.4) \times (11.5) \times (\text{cm}) \times (\text{cm})$
    $A = 96.6 \text{ cm}^2$

77. We draw a picture of the situation.

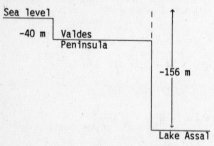

To find how much lower Lake Assal is, we subtract:

   $-40 - (-156) = -40 + 156 = 116$

Lake Assal is 116 m lower than the Valdes Peninsula.

Exercise Set 11.4

1. -16

3. -24

5. -72

7. 16

9. 42

11. -120

13. -238

15. 1200

17. 98

19. -12.4

21. 24

23. 21.7

25. $\frac{2}{3} \cdot \left(-\frac{3}{5}\right) = -\left(\frac{2 \cdot 3}{3 \cdot 5}\right) = -\left(\frac{2}{5} \cdot \frac{3}{3}\right) = -\frac{2}{5}$

27. $-\frac{3}{8} \cdot \left(-\frac{2}{9}\right) = \frac{3 \cdot 2 \cdot 1}{4 \cdot 2 \cdot 3 \cdot 3} = \frac{1}{12}$

29. -17.01

31. $-\frac{5}{9} \cdot \frac{3}{4} = -\frac{5 \cdot 3}{3 \cdot 3 \cdot 4} = -\frac{5}{12}$

33. $7 \cdot (-4) \cdot (-3) \cdot 5 = 7 \cdot 12 \cdot 5 = 7 \cdot 60 = 420$

35. $-\frac{2}{3} \cdot \frac{1}{2} \cdot \left(-\frac{6}{7}\right) = -\frac{2}{6} \cdot \left(-\frac{6}{7}\right) = \frac{2 \cdot 6}{7 \cdot 6} = \frac{2}{7}$

37. $-3 \cdot (-4) \cdot (-5) = 12 \cdot (-5) = -60$

39. $-2 \cdot (-5) \cdot (-3) \cdot (-5) = 10 \cdot 15 = 150$

41. $-\frac{2}{45}$

43. $-7 \cdot (-21) \cdot 13 = 147 \cdot 13 = 1911$

45. $-4 \cdot (-1.8) \cdot 7 = (7.2) \cdot 7 = 50.4$

47. $-\frac{1}{9} \cdot \left(-\frac{2}{3}\right) \cdot \left(\frac{5}{7}\right) = \frac{2}{27} \cdot \frac{5}{7} = \frac{10}{189}$

49. $4 \cdot (-4) \cdot (-5) \cdot (-12) = -16 \cdot (60) = -960$

51. $0.07 \cdot (-7) \cdot 6 \cdot (-6) = 0.07 \cdot 6 \cdot (-7) \cdot (-6) =$
    $0.42 \cdot (42) = 17.64$

53. $\left(-\frac{5}{6}\right)\left(\frac{1}{8}\right)\left(-\frac{3}{7}\right)\left(-\frac{1}{7}\right) = \left(-\frac{5}{48}\right)\left(\frac{3}{49}\right) = -\frac{5 \cdot 3}{16 \cdot 3 \cdot 49} =$
    $-\frac{5}{784}$

55. $(-14)\cdot(-27)\cdot(-2) = 378\cdot(-2) = -756$

57. $(-8)(-9)(-10) = 72(-10) = -720$

59. $(-6)(-7)(-8)(-9)(-10) = 42\cdot72\cdot(-10) =$
$3024\cdot(-10) = -30,240$

61.
```
          3
       3│9
      2│1 8
      2│3 6
      2│7 2
    2│1 4 4
    2│2 8 8
    2│5 7 6
  2│1 1 5 2
  2│2 3 0 4
  2│4 6 0 8
```

$4608 = 2\cdot2\cdot2\cdot2\cdot2\cdot2\cdot2\cdot2\cdot3\cdot3$

63. Using an equation:

Translate: 23 is what percent, of 69?
             ↓ ↓   ↓   ↓ ↓
             23 =   n   × 69

To solve the equation we divide on both sides
by 69 and convert to percent notation.

$$n\cdot69 = 23$$
$$\frac{n\cdot69}{69} = \frac{23}{69}$$
$$n = 0.33\tfrac{1}{3}, \text{ or } 0.33\overline{3} = 33\tfrac{1}{3}\%, \text{ or } 33.\overline{3}\%$$

Thus, 23 is $33\tfrac{1}{3}\%$, or $33.\overline{3}\%$, of 69.

Using a proportion:

23 is what percent of 69?   Percents   Quantities
 ↓   ↓   ↓   0% ──────────── 0
amount   numbers   base   n% ──────────── 23

                              100% ──────────── 69

Translate: $\dfrac{n}{100} = \dfrac{23}{69}$

Solve: $69\cdot n = 100\cdot23$ Finding cross-products

$n = \dfrac{100\cdot23}{69}$ Dividing by 69

$n = \dfrac{100\cdot23}{3\cdot23}$

$n = \dfrac{100}{3} = 33\tfrac{1}{3}$, or $33.\overline{3}$

Thus, 23 is $33\tfrac{1}{3}\%$, or $33.\overline{3}\%$, of 69.

65. $-6[(-5) + (-7)] = -6[-12] = 72$

67. $-(3^5)\cdot[-(2^3)] = -243[-8] = 1944$

69. $|(-2)^3 + 4^2| - (2 - 7)^2 = |(-2)^3 + 4^2| - (-5)^2 =$
$|-8 + 16| - 25 = |8| - 25 = 8 - 25 = -17$

71. a) m and n have different signs;

b) either m or n is zero;

c) m and n have the same sign

Exercise Set 11.5

---

1. $36 \div (-6) = -6$ Check: $-6\cdot(-6) = 36$

3. $\dfrac{26}{-2} = -13$ Check: $-13\cdot(-2) = 26$

5. $\dfrac{-16}{8} = -2$ Check: $-2\cdot8 = -16$

7. $\dfrac{-48}{-12} = 4$ Check: $4(-12) = -48$

9. $\dfrac{-72}{9} = -8$ Check: $-8\cdot9 = -72$

11. $-100 \div (-50) = 2$ Check: $2(-50) = -100$

13. $-108 \div 9 = -12$ Check: $9(-12) = -108$

15. $\dfrac{200}{-25} = -8$ Check: $-8(-25) = 200$

17. Undefined

19. $\dfrac{81}{-9} = -9$ Check: $-9\cdot(-9) = 81$

21. The reciprocal of $-\dfrac{15}{7}$ is $-\dfrac{7}{15}$ because
$-\dfrac{15}{7} \cdot \left(-\dfrac{7}{15}\right) = 1.$

23. The reciprocal of 13 is $\dfrac{1}{13}$ because $13 \cdot \dfrac{1}{13} = 1.$

25. $\dfrac{3}{4} \div \left(-\dfrac{2}{3}\right) = \dfrac{3}{4} \cdot \left(-\dfrac{3}{2}\right) = -\dfrac{9}{8}$

27. $-\dfrac{5}{4} \div \left(-\dfrac{3}{4}\right) = -\dfrac{5}{4} \cdot \left(-\dfrac{4}{3}\right) = \dfrac{20}{12} = \dfrac{5\cdot4}{3\cdot4} = \dfrac{5}{3}$

29. $-\dfrac{2}{7} \div \left(-\dfrac{4}{9}\right) = -\dfrac{2}{7} \cdot \left(-\dfrac{9}{4}\right) = \dfrac{18}{28} = \dfrac{9\cdot2}{14\cdot2} = \dfrac{9}{14}$

31. $-\dfrac{3}{8} \div \left(-\dfrac{8}{3}\right) = -\dfrac{3}{8} \cdot \left(-\dfrac{3}{8}\right) = \dfrac{9}{64}$

33. $-6.6 \div 3.3 = -2$ Do the long division. Make
the answer negative.

35. $\dfrac{-11}{-13} = \dfrac{11}{13}$ The opposite of a number divided by
the opposite of another number is
the quotient of the two numbers.

37. $\dfrac{48.6}{-3} = -16.2$ Do the long division. Make the
answer negative.

39. $\dfrac{-9}{17 - 17} = \dfrac{-9}{0}$

Division by zero is undefined.

41. $8 - 2 \cdot 3 - 9 = 8 - 6 - 9$ Multiplying

$\qquad = 2 - 9$ Doing all additions and
$\qquad = -7$ subtractions in order
$\qquad\qquad\qquad\qquad$ from left to right

43. $(8 - 2 \cdot 3) - 9 = (8 - 6) - 9$ Multiplying inside
$\qquad\qquad\qquad\qquad\qquad\qquad$ parentheses

$\qquad\qquad\quad = 2 - 9$ Subtracting inside
$\qquad\qquad\qquad\qquad\qquad$ parentheses

$\qquad\qquad\quad = -7$ Subtracting

45. $16 \cdot (-24) + 50 = -384 + 50$ Multiplying
$\qquad\qquad\qquad\quad = -334$ Adding

47. $2^4 + 2^3 - 10$

$\quad = 16 + 8 - 10$ Evaluating exponential expressions
$\quad = 24 - 10$ Adding and subtracting in order
$\quad = 14$ from left to right

49. $5^3 + 26 \cdot 71 - (16 + 25 \cdot 3)$

$\quad = 5^3 + 26 \cdot 71 - (16 + 75)$ Multiplying inside
$\qquad\qquad\qquad\qquad\qquad\qquad\quad$ parentheses
$\quad = 5^3 + 26 \cdot 71 - 91$ Adding inside parentheses
$\quad = 125 + 26 \cdot 71 - 91$ Evaluating the exponential
$\qquad\qquad\qquad\qquad\qquad$ expression
$\quad = 125 + 1846 - 91$ Multiplying
$\quad = 1971 - 91$ Adding and subtracting in
$\quad = 1880$ order from left to right

51. $4 \cdot 5 - 2 \cdot 6 + 4$

$\quad = 20 - 12 + 4$ Doing the multiplications
$\quad = 8 + 4$
$\quad = 12$

53. $\dfrac{4^3}{8} = \dfrac{64}{8}$ Evaluating the exponential expression

$\qquad = 8$ Dividing

55. $8(-7) + 6(-5) = -56 - 30$ Multiplying
$\qquad\qquad\qquad\quad = -86$

57. $19 - 5(-3) + 3 = 19 + 15 + 3$ Multiplying
$\qquad\qquad\qquad\qquad = 34 + 3$
$\qquad\qquad\qquad\qquad = 37$

59. $9 \div (-3) + 16 \div 8 = -3 + 2$ Dividing
$\qquad\qquad\qquad\qquad\qquad = -1$

61. $6 - 4^2 = 6 - 16$
$\qquad\quad = -10$

63. $(3 - 8)^2 = (-5)^2$ Subtracting inside parentheses
$\qquad\quad = 25$

65. $12 - 20^3 = 12 - 8000$
$\qquad\qquad = -7988$

67. $2 \times 10^3 - 5000 = 2 \times 1000 - 5000$

$\qquad\qquad\qquad = 2000 - 5000$

$\qquad\qquad\qquad = -3000$

69. $6[9 - (3 - 4)]$

$= 6[9 - (-1)]$ Subtracting inside the innermost
$\qquad\qquad\qquad$ parentheses
$= 6[9 + 1]$
$= 6[10]$
$= 60$

71. $-1000 \div (-100) \div 10$

$= 10 \div 10$ Doing the divisions in order
$= 1$ from left to right

73. $8 - (7 - 9) = 8 - (-2)$
$\qquad\qquad = 8 + 2$
$\qquad\qquad = 10$

75. $\dfrac{10 - 6^2}{9^2 + 3^2} = \dfrac{10 - 36}{81 + 9}$ Evaluating the exponential
$\qquad\qquad\qquad\qquad\qquad$ expressions

$\qquad = \dfrac{-26}{90}$ Subtracting in the numerator,
$\qquad\qquad\qquad$ adding in the denominator

$\qquad = -\dfrac{13}{45}$ Simplifying

77. $\dfrac{20(8 - 3) - 4(10 - 3)}{10(2 - 6) - 2(5 + 2)} = \dfrac{20 \cdot 5 - 4 \cdot 7}{10(-4) - 2 \cdot 7}$ Doing the
$\qquad\qquad\qquad\qquad\qquad\qquad\qquad\qquad\qquad$ operations
$\qquad\qquad\qquad\qquad\qquad\qquad\qquad\qquad\qquad$ inside
$\qquad\qquad\qquad\qquad\qquad\qquad\qquad\qquad\qquad$ parentheses

$\qquad\qquad\qquad = \dfrac{100 - 28}{-40 - 14}$ Multiplying

$\qquad\qquad\qquad = \dfrac{72}{-54}$ Subtracting

$\qquad\qquad\qquad = -\dfrac{4}{3}$ Simplifying

Exercise Set 12.1

1. $6x = 6 \cdot 7 = 42$

3. $\frac{x}{y} = \frac{9}{3} = 3$

5. $\frac{3p}{q} = \frac{3 \cdot 2}{6} = \frac{6}{6} = 1$

7. $\frac{x + y}{5} = \frac{10 + 20}{5} = \frac{30}{5} = 6$

9. $10(x + y) = 10(20 + 4) = 10 \cdot 24 = 240$
   $10x + 10y = 10 \cdot 20 + 10 \cdot 4 = 200 + 40 = 240$

11. $10(x - y) = 10(20 - 4) = 10 \cdot 16 = 160$
    $10x - 10y = 10 \cdot 20 - 10 \cdot 4 = 200 - 40 = 160$

13. $2(b + 5) = 2 \cdot b + 2 \cdot 5 = 2b + 10$

15. $7(1 - t) = 7 \cdot 1 - 7 \cdot t = 7 - 7t$

17. $6(5x + 2) = 6 \cdot 5x + 6 \cdot 2 = 30x + 12$

19. $7(x + 4 + 6y) = 7 \cdot x + 7 \cdot 4 + 7 \cdot 6y = 7x + 28 + 42y$

21. $-7(y - 2) = -7 \cdot y - (-7) \cdot 2 = -7y - (-14) =$
    $-7y + 14$

23. $-9(-5x - 6y + 8) = -9(-5x) - (-9)6y + (-9)8 =$
    $45x - (-54)y + (-72) = 45x + 54y - 72$

25. $-4(x - 3y - 2z) = -4 \cdot x - (-4)3y - (-4)2z =$
    $-4x - (-12)y - (-8)z = -4x + 12y + 8z$

27. $3.1(-1.2x + 3.2y - 1.1) = 3.1(-1.2x) +$
    $(3.1)3.2y - 3.1(1.1) = -3.72x + 9.92y - 3.41$

29. $2x + 4 = 2 \cdot x + 2 \cdot 2 = 2(x + 2)$

31. $30 + 5y = 5 \cdot 6 + 5 \cdot y = 5(6 + y)$

33. $14x + 21y = 7 \cdot 2x + 7 \cdot 3y = 7(2x + 3y)$

35. $5x + 10 + 15y = 5 \cdot x + 5 \cdot 2 + 5 \cdot 3y = 5(x + 2 + 3y)$

37. $8x - 24 = 8 \cdot x - 8 \cdot 3 = 8(x - 3)$

39. $32 - 4y = 4 \cdot 8 - 4 \cdot y = 4(8 - y)$

41. $8x + 10y - 22 = 2 \cdot 4x + 2 \cdot 5y - 2 \cdot 11 =$
    $2(4x + 5y - 11)$

43. $18x - 12y + 6 = 6 \cdot 3x - 6 \cdot 2y + 6 \cdot 1 =$
    $6(3x - 2y + 1)$

45. $9a + 10a = (9 + 10)a = 19a$

47. $10a - a = 10a - 1 \cdot a = (10 - 1)a = 9a$

49. $2x + 9z + 6x$
    $= 2x + 6x + 9z$
    $= (2 + 6)x + 9z$
    $= 8x + 9z$

51. $41a + 90 - 60a - 2$
    $= 41a - 60a + 90 - 2$
    $= (41 - 60)a + (90 - 2)$
    $= -19a + 88$

53. $23 + 5t + 7y - t - y - 27$
    $= 23 - 27 + 5t - 1 \cdot t + 7y - 1 \cdot y$
    $= (23 - 27) + (5 - 1)t + (7 - 1)y$
    $= -4 + 4t + 6y, \text{ or } 4t + 6y - 4$

55. $11x - 3x = (11 - 3)x = 8x$

57. $6n - n = (6 - 1)n = 5n$

59. $y - 17y = (1 - 17)y = -16y$

61. $-8 + 11a - 5b + 6a - 7b + 7 =$
    $11a + 6a - 5b - 7b - 8 + 7 =$
    $(11 + 6)a + (-5 - 7)b + (-8 + 7) = 17a - 12b - 1$

63. $9x + 2y - 5x = (9 - 5)x + 2y = 4x + 2y$

65. $11x + 2y - 4x - y = (11 - 4)x + (2 - 1)y = 7x + y$

67. $2.7x + 2.3y - 1.9x - 1.8y = (2.7 - 1.9)x +$
    $(2.3 - 1.8)y = 0.8x + 0.5y$

Exercise Set 12.2

1. $\quad x + 2 = 6$
   $x + 2 - 2 = 6 - 2 \quad$ Subtracting 2 on both sides
   $\quad x = 4 \qquad$ Simplifying

   Check: $\dfrac{x + 2 = 6}{\begin{array}{c|c} 4 + 2 & 6 \\ 6 & \end{array}}$ TRUE

3. $\quad x + 15 = -5$
   $x + 15 - 15 = -5 - 15 \quad$ Subtracting 15 on both sides
   $\quad x = -20$

   Check: $\dfrac{x + 15 = -5}{\begin{array}{c|c} -20 + 15 & -5 \\ -5 & \end{array}}$ TRUE

5. $\quad x + 6 = -8 \qquad$ Check: $\dfrac{x + 6 = -8}{\begin{array}{c|c} -14 + 6 & -8 \\ -8 & \end{array}}$ TRUE
   $x + 6 - 6 = -8 - 6$
   $\quad x = -14$

7.
$$x + 16 = -2$$
$$x + 16 - 16 = -2 - 16$$
$$x = -18$$

Check:
$$\frac{x + 16 = -2}{-18 + 16 \mid -2}$$
$$-2 \mid \text{TRUE}$$

9.
$$x - 9 = 6$$
$$x - 9 + 9 = 6 + 9$$
$$x = 15$$

Check:
$$\frac{x - 9 = 6}{15 - 9 \mid 6}$$
$$6 \mid \text{TRUE}$$

11.
$$x - 7 = -21$$
$$x - 7 + 7 = -21 + 7$$
$$x = -14$$

Check:
$$\frac{x - 7 = -21}{-14 - 7 \mid -21}$$
$$-21 \mid \text{TRUE}$$

13.
$$5 + t = 7$$
$$-5 + 5 + t = -5 + 7$$
$$t = 2$$

Check:
$$\frac{5 + t = 7}{5 + 2 \mid 7}$$
$$7 \mid \text{TRUE}$$

15.
$$-7 + y = 13$$
$$7 + (-7) + y = 7 + 13$$
$$y = 20$$

Check:
$$\frac{-7 + y = 13}{-7 + 20 \mid 13}$$
$$13 \mid \text{TRUE}$$

17.
$$-3 + t = -9$$
$$3 + (-3) + t = 3 + (-9)$$
$$t = -6$$

Check:
$$\frac{-3 + t = -9}{-3 + (-6) \mid -9}$$
$$-9 \mid \text{TRUE}$$

19.
$$r + \frac{1}{3} = \frac{8}{3}$$
$$r + \frac{1}{3} - \frac{1}{3} = \frac{8}{3} - \frac{1}{3}$$
$$r = \frac{7}{3}$$

Check:
$$\frac{r + \frac{1}{3} = \frac{8}{3}}{\frac{7}{3} + \frac{1}{3} \mid \frac{8}{3}}$$
$$\frac{8}{3} \mid \text{TRUE}$$

21.
$$m + \frac{5}{6} = -\frac{11}{12}$$
$$m + \frac{5}{6} - \frac{5}{6} = -\frac{11}{12} - \frac{5}{6}$$
$$m = -\frac{11}{12} - \frac{5}{6}\left(\frac{2}{2}\right)$$
$$m = -\frac{11}{12} - \frac{10}{12}$$
$$m = -\frac{21}{12} = -\frac{3 \cdot 7}{3 \cdot 4}$$
$$m = -\frac{7}{4}$$

Check:
$$\frac{m + \frac{5}{6} = -\frac{11}{12}}{-\frac{7}{4} + \frac{5}{6} \mid -\frac{11}{12}}$$
$$-\frac{21}{12} + \frac{10}{12}$$
$$-\frac{11}{12} \mid \text{TRUE}$$

23.
$$x - \frac{5}{6} = \frac{7}{8}$$
$$x - \frac{5}{6} + \frac{5}{6} = \frac{7}{8} + \frac{5}{6}$$
$$x = \frac{7}{8} \cdot \frac{3}{3} + \frac{5}{6} \cdot \frac{4}{4}$$
$$x = \frac{21}{24} + \frac{20}{24}$$
$$x = \frac{41}{24}$$

Check:
$$\frac{x - \frac{5}{6} = \frac{7}{8}}{\frac{41}{24} - \frac{5}{6} \mid \frac{7}{8}}$$
$$\frac{41}{24} - \frac{20}{24} \mid \frac{21}{24}$$
$$\frac{21}{24} \mid \text{TRUE}$$

25.
$$-\frac{1}{5} + z = -\frac{1}{4}$$
$$\frac{1}{5} - \frac{1}{5} + z = \frac{1}{5} - \frac{1}{4}$$
$$z = \frac{1}{5} \cdot \frac{4}{4} - \frac{1}{4} \cdot \frac{5}{5}$$
$$z = \frac{4}{20} - \frac{5}{20}$$
$$z = -\frac{1}{20}$$

Check:
$$\frac{-\frac{1}{5} + z = -\frac{1}{4}}{-\frac{1}{5} + \left(-\frac{1}{20}\right) \mid -\frac{1}{4}}$$
$$-\frac{4}{20} + \left(-\frac{1}{20}\right) \mid -\frac{5}{20}$$
$$-\frac{5}{20} \mid \text{TRUE}$$

27.
$$x + 2.3 = 7.4$$
$$x + 2.3 - 2.3 = 7.4 - 2.3$$
$$x = 5.1$$

Check:
$$\frac{x + 2.3 = 7.4}{5.1 + 2.3 \mid 7.4}$$
$$7.4 \mid \text{TRUE}$$

29.
$$7.6 = x - 4.8$$
$$7.6 + 4.8 = x - 4.8 + 4.8$$
$$12.4 = x$$

Check:
$$\frac{7.6 = x - 4.8}{7.6 \mid 12.4 - 4.8}$$
$$\mid 7.6 \quad \text{TRUE}$$

31.
$$-9.7 = -4.7 + y$$
$$4.7 + (-9.7) = 4.7 + (-4.7) + y$$
$$-5 = y$$

Check:
$$\frac{-9.7 = -4.7 + y}{-9.7 \mid -4.7 + (-5)}$$
$$\mid -9.7 \quad \text{TRUE}$$

33.
$$5\frac{1}{6} + x = 7$$
$$-5\frac{1}{6} + 5\frac{1}{6} + x = -5\frac{1}{6} + 7$$
$$x = -\frac{31}{6} + \frac{42}{6}$$
$$x = \frac{11}{6}, \text{ or } 1\frac{5}{6}$$

Check:
$$\frac{5\frac{1}{6} + x = 7}{5\frac{1}{6} + 1\frac{5}{6} \mid 7}$$
$$7 \mid \text{TRUE}$$

35. $q + \frac{1}{3} = -\frac{1}{7}$

$q + \frac{1}{3} - \frac{1}{3} = -\frac{1}{7} - \frac{1}{3}$

$q = -\frac{1}{7} \cdot \frac{3}{3} - \frac{1}{3}\left(\frac{7}{7}\right)$

$q = -\frac{3}{21} - \frac{7}{21}$

$q = -\frac{10}{21}$

Check:

$$\begin{array}{c|c} q + \frac{1}{3} = -\frac{1}{7} \\ \hline -\frac{10}{21} + \frac{1}{3} & -\frac{1}{7} \\ -\frac{10}{21} + \frac{7}{21} & -\frac{3}{21} \\ -\frac{3}{21} & \text{TRUE} \end{array}$$

37. $-3 + (-8)$ Two negative numbers. We add the absolute values, getting 11, and make the answer negative. $-3 + (-8) = -11$

39. $-\frac{2}{3} \cdot \frac{5}{8} = -\frac{2 \cdot 5}{3 \cdot 8} = -\frac{2 \cdot 5}{3 \cdot 2 \cdot 4} = -\frac{5}{12}$

41.

$-356.788 = -699.034 + t$

$699.034 + (-356.788) = 699.034 + (-699.034) + t$

$342.246 = t$

43. $x + \frac{4}{5} = -\frac{2}{3} - \frac{4}{15}$

$x + \frac{4}{5} - \frac{4}{5} = -\frac{2}{3} - \frac{4}{15} - \frac{4}{5}$

$x = -\frac{2}{3} \cdot \frac{5}{5} - \frac{4}{15} - \frac{4}{5} \cdot \frac{3}{3}$

$x = -\frac{10}{15} - \frac{4}{15} - \frac{12}{15}$

$x = -\frac{26}{15}$

45. $16 + x - 22 = -16$

$x - 6 = -16$ Adding on the left side

$x - 6 + 6 = -16 + 6$

$x = -10$

47. $x + 3 = 3 + x$

$x + 3 - 3 = 3 + x - 3$

$x = x$

$x = x$ is true for all real numbers. Thus the solution is all real numbers.

49. $-\frac{3}{2} + x = -\frac{5}{17} - \frac{3}{2}$

$\frac{3}{2} - \frac{3}{2} + x = \frac{3}{2} - \frac{5}{17} - \frac{3}{2}$

$x = \left(\frac{3}{2} - \frac{3}{2}\right) - \frac{5}{17}$

$x = -\frac{5}{17}$

51. $|x| + 6 = 19$

$|x| = 13$

x represents a number whose distance from 0 is 13. Thus, x = -13 or x = 13.

Exercise Set 12.3

1. $6x = 36$

$\frac{6x}{6} = \frac{36}{6}$

$1 \cdot x = 6$

$x = 6$

Check:
$$\begin{array}{c|c} 6x = 36 \\ \hline 6 \cdot 6 & 36 \\ 36 & \text{TRUE} \end{array}$$

3. $5x = 45$

$\frac{5x}{5} = \frac{45}{5}$

$1 \cdot x = 9$

$x = 9$

Check:
$$\begin{array}{c|c} 5x = 45 \\ \hline 5 \cdot 9 & 45 \\ 45 & \text{TRUE} \end{array}$$

5. $84 = 7x$

$\frac{84}{7} = \frac{7x}{7}$

$12 = x$

Check:
$$\begin{array}{c|c} 84 = 7x \\ \hline 84 & 7 \cdot 12 \\ & 84 \quad \text{TRUE} \end{array}$$

7. $-x = 40$

$-1 \cdot x = 40$

$-1 \cdot (-1 \cdot x) = -1 \cdot 40$

$1 \cdot x = -40$

$x = -40$

Check:
$$\begin{array}{c|c} -x = 40 \\ \hline -(-40) & 40 \\ 40 & \text{TRUE} \end{array}$$

9. $-2x = -10$

$\frac{-2x}{-2} = \frac{-10}{-2}$

$x = 5$

Check:
$$\begin{array}{c|c} -2x = -10 \\ \hline -2 \cdot 5 & -10 \\ -10 & \text{TRUE} \end{array}$$

11. $7x = -49$

$\frac{7x}{7} = \frac{-49}{7}$

$x = -\frac{49}{7}$

$x = -7$

Check:
$$\begin{array}{c|c} 7x = -49 \\ \hline 7(-7) & -49 \\ -49 & \text{TRUE} \end{array}$$

13. $-12x = 72$

$\frac{-12x}{-12} = \frac{72}{-12}$

$x = -\frac{72}{12}$

$x = -6$

Check:
$$\begin{array}{c|c} -12x = 72 \\ \hline -12(-6) & 72 \\ 72 & \text{TRUE} \end{array}$$

15. $-21x = -126$

$\frac{-21x}{-21} = \frac{-126}{-21}$

$x = \frac{126}{21}$

$x = 6$

Check:
$$\begin{array}{c|c} -21x = -126 \\ \hline -21 \cdot 6 & 126 \\ -126 & \text{TRUE} \end{array}$$

17. $\frac{1}{7}t = -9$ Check: $\frac{1}{7}t = -9$

    $7 \cdot \left[\frac{1}{7}t\right] = 7 \cdot (-9)$ $\dfrac{\frac{1}{7} \cdot (-63)}{} \,\Big|\, -9$

       $t = -63$ -9 TRUE

19. $\frac{3}{4}x = 27$ Check: $\frac{3}{4}x = 27$

    $\frac{4}{3} \cdot \frac{3}{4}x = \frac{4}{3} \cdot 27$ $\dfrac{\frac{3}{4} \cdot 36}{} \,\Big|\, 27$

       $x = \frac{4 \cdot 3 \cdot 3}{3 \cdot 1}$ 27 TRUE

       $x = 36$

21. $-\frac{1}{3}t = 7$ Check: $-\frac{1}{3}t = 7$

    $-3 \cdot \left[-\frac{1}{3}\right] \cdot t = -3 \cdot 7$ $\dfrac{-\frac{1}{3} \cdot (-21)}{} \,\Big|\, 7$

       $t = -21$ 7 TRUE

23. $-\frac{1}{3}m = \frac{1}{5}$ Check: $-\frac{1}{3}m = \frac{1}{5}$

    $-3 \cdot \left[-\frac{1}{3} \cdot m\right] = -3 \cdot \frac{1}{5}$ $\dfrac{-\frac{1}{3} \cdot \left[-\frac{3}{5}\right]}{} \,\Big|\, \frac{1}{5}$

       $m = -\frac{3}{5}$ $\frac{1}{5}$ TRUE

25. $-\frac{3}{5}r = \frac{9}{10}$ Check: $-\frac{3}{5}r = \frac{9}{10}$

    $-\frac{5}{3} \cdot \left[-\frac{3}{5}r\right] = -\frac{5}{3} \cdot \frac{9}{10}$ $\dfrac{-\frac{3}{5} \cdot \left[-\frac{3}{2}\right]}{} \,\Big|\, \frac{9}{10}$

       $r = -\frac{5 \cdot 3 \cdot 3}{3 \cdot 5 \cdot 2}$ $\frac{9}{10}$ TRUE

       $r = -\frac{3}{2}$

27. $-\frac{3}{2}r = -\frac{27}{4}$ Check: $-\frac{3}{2}r = -\frac{27}{4}$

    $-\frac{2}{3} \cdot \left[-\frac{3}{2}r\right] = -\frac{2}{3} \cdot \left[-\frac{27}{4}\right]$ $\dfrac{-\frac{3}{2} \cdot \frac{9}{2}}{} \,\Big|\, -\frac{27}{4}$

       $r = \frac{2 \cdot 3 \cdot 3}{3 \cdot 2 \cdot 2}$ $-\frac{27}{4}$ TRUE

       $r = \frac{9}{2}$

29. $6.3x = 44.1$ Check: $6.3x = 44.1$

    $\frac{6.3x}{6.3} = \frac{44.1}{6.3}$ $\dfrac{6.3 \cdot 7}{} \,\Big|\, 44.1$

       $x = \frac{44.1}{6.3}$ 44.1 TRUE

       $x = 7$

31. $-3.1y = 21.7$

    $\frac{-3.1y}{-3.1} = \frac{21.7}{-3.1}$

       $y = -\frac{21.7}{3.1}$

       $y = -7$

    Check: $-3.1y = 21.7$

    $\dfrac{-3.1(-7)}{} \,\Big|\, 21.7$

       21.7 TRUE

33. $38.7m = 309.6$

    $\frac{38.7m}{38.7} = \frac{309.6}{38.7}$

       $m = \frac{309.6}{38.7}$

       $m = 8$

    Check: $\frac{38.7m = 309.6}{38.7 \cdot 8 \,\Big|\, 309.6}$

       309.6 TRUE

35. $-\frac{2}{3}y = -10.6$

    $-\frac{3}{2} \cdot \left[-\frac{2}{3}y\right] = -\frac{3}{2} \cdot (-10.6)$

       $y = \frac{31.8}{2}$

       $y = 15.9$

    Check: $-\frac{2}{3}y = -10.6$

    $\dfrac{-\frac{2}{3}(15.9)}{} \,\Big|\, -10.6$

    $-\frac{31.8}{3}$

       -10.6 TRUE

37. $C = 2 \cdot \pi \cdot r$

    $C \approx 2 \cdot 3.14 \cdot 10 \text{ ft} = 62.8 \text{ ft}$

    $d = 2 \cdot r$

    $d = 2 \cdot 10 \text{ ft} = 20 \text{ ft}$

    $A = \pi \cdot r \cdot r$

    $A \approx 3.14 \cdot 10 \text{ ft} \cdot 10 \text{ ft} = 314 \text{ ft}^2$

39. $V = \ell \cdot w \cdot h$

    $V = 25 \text{ ft} \cdot 10 \text{ ft} \cdot 32 \text{ ft} = 8000 \text{ ft}^3$

41. $-0.2344m = 2028.732$

    $\frac{-0.2344m}{-0.2344} = \frac{2028.732}{-0.2344}$

       $m = -\frac{2028.732}{0.2344}$

       $m = -8655$

43. For all x, $0 \cdot x = 0$. There is no solution to $0 \cdot x = 9$.

45. $2|x| = -12$

    $\frac{2|x|}{2} = \frac{-12}{2}$

    $|x| = -6$

Absolute value cannot be negative. The equation has no solution.

Exercise Set 12.4

1. $5x + 6 = 31$
    $5x + 6 - 6 = 31 - 6$
    $5x = 25$
    $\dfrac{5x}{5} = \dfrac{25}{5}$
    $x = 5$

    Check: $\dfrac{5x + 6 = 31}{5 \cdot 5 + 6 \,\Big|\, 31}$
    $25 + 6$
    $31 \,\Big|\,$ TRUE

3. $8x + 4 = 68$
    $8x + 4 - 4 = 68 - 4$
    $8x = 64$
    $\dfrac{8x}{8} = \dfrac{64}{8}$
    $x = 8$

    Check: $\dfrac{8x + 4 = 68}{8 \cdot 8 + 4 \,\Big|\, 68}$
    $64 + 4$
    $68 \,\Big|\,$ TRUE

5. $4x - 6 = 34$
    $4x - 6 + 6 = 34 + 6$
    $4x = 40$
    $\dfrac{4x}{4} = \dfrac{40}{4}$
    $x = 10$

    Check: $\dfrac{4x - 6 = 34}{4 \cdot 10 - 6 \,\Big|\, 34}$
    $40 - 6$
    $34 \,\Big|\,$ TRUE

7. $3x - 9 = 33$
    $3x - 9 + 9 = 33 + 9$
    $3x = 42$
    $\dfrac{3x}{3} = \dfrac{42}{3}$
    $x = 14$

    Check: $\dfrac{3x - 9 = 33}{3 \cdot 14 - 9 \,\Big|\, 33}$
    $42 - 9$
    $33 \,\Big|\,$ TRUE

9. $7x + 2 = -54$
    $7x + 2 - 2 = -54 - 2$
    $7x = -56$
    $\dfrac{7x}{7} = \dfrac{-56}{7}$
    $x = -8$

    Check: $\dfrac{7x + 2 = -54}{7(-8) + 2 \,\Big|\, -54}$
    $-56 + 2$
    $-54 \,\Big|\,$ TRUE

11. $-45 = 6y + 3$
    $-45 - 3 = 6y + 3 - 3$
    $-48 = 6y$
    $-\dfrac{48}{6} = \dfrac{6y}{6}$
    $-8 = y$

    Check: $\dfrac{-45 = 6y + 3}{-45 \,\Big|\, 6(-8) + 3}$
    $-48 + 3$
    $-45 \,\Big|\,$ TRUE

13. $-4x + 7 = 35$
    $-4x + 7 - 7 = 35 - 7$
    $-4x = 28$
    $\dfrac{4x}{-4} = \dfrac{28}{-4}$
    $x = -7$

    Check: $\dfrac{-4x + 7 = 35}{-4(-7) + 7 \,\Big|\, 35}$
    $28 + 7$
    $35 \,\Big|\,$ TRUE

15. $-7x - 24 = -129$
    $-7x - 24 + 24 = -129 + 24$
    $-7x = -105$
    $\dfrac{-7x}{-7} = \dfrac{-105}{-7}$
    $x = 15$

    Check: $\dfrac{-7x - 24 = -129}{-7 \cdot 15 - 24 \,\Big|\, -129}$
    $-105 - 24$
    $-129 \,\Big|\,$ TRUE

17. $5x + 7x = 72$
    $12x = 72$
    $\dfrac{12x}{12} = \dfrac{72}{12}$
    $x = 6$

    Check: $\dfrac{5x + 7x = 72}{5 \cdot 6 + 7 \cdot 6 \,\Big|\, 72}$
    $30 + 42$
    $72 \,\Big|\,$ TRUE

19. $8x + 7x = 60$
    $15x = 60$
    $\dfrac{15x}{15} = \dfrac{60}{15}$
    $x = 4$

    Check: $\dfrac{8x + 7x = 60}{8 \cdot 4 + 7 \cdot 4 \,\Big|\, 60}$
    $32 + 28$
    $60 \,\Big|\,$ TRUE

21. $4x + 3x = 42$
    $7x = 42$
    $\dfrac{7x}{7} = \dfrac{42}{7}$
    $x = 6$

    Check: $\dfrac{4x + 3x = 42}{4 \cdot 6 + 3 \cdot 6 \,\Big|\, 42}$
    $24 + 18$
    $42 \,\Big|\,$ TRUE

23. $-6y - 3y = 27$
    $-9y = 27$
    $\dfrac{-9y}{-9} = \dfrac{27}{-9}$
    $y = -3$

    Check: $\dfrac{-6y - 3y = 27}{-6(-3) - 3(-3) \,\Big|\, 27}$
    $18 + 9$
    $27 \,\Big|\,$ TRUE

25. $-7y - 8y = -15$
    $-15y = -15$
    $\dfrac{-15y}{-15} = \dfrac{-15}{-15}$
    $y = 1$

    Check: $\dfrac{-7y - 8y = -15}{-7 \cdot 1 - 8 \cdot 1 \,\Big|\, -15}$
    $-7 - 8$
    $-15 \,\Big|\,$ TRUE

27. $10.2y - 7.3y = -58$
    $2.9y = -58$
    $\dfrac{2.9y}{2.9} = \dfrac{-58}{2.9}$
    $y = -\dfrac{58}{2.9}$
    $y = -20$

    Check: $\dfrac{10.2y - 7.3y = -58}{10.2(-20) - 7.3(-20) \,\Big|\, -58}$
    $-204 + 146$
    $-58 \,\Big|\,$ TRUE

29. $x + \frac{1}{3}x = 8$ Check: $\dfrac{x + \frac{1}{3}x = 8}{6 + \frac{1}{3} \cdot 6 \mid 8}$

$\left(1 + \frac{1}{3}\right)x = 8$

$\frac{4}{3}x = 8$ $6 + 2$

$\frac{3}{4} \cdot \frac{4}{3}x = \frac{3}{4} \cdot 8$ $8 \mid$ TRUE

$x = 6$

31. $8y - 35 = 3y$ Check: $\dfrac{8y - 35 = 3y}{8 \cdot 7 - 35 \mid 3 \cdot 7}$

$8y = 3y + 35$

$8y - 3y = 35$ $56 - 35 \mid 21$

$5y = 35$ $21 \mid$ TRUE

$y = \frac{35}{5}$

$y = 7$

33. $8x - 1 = 23 - 4x$ Check: $\dfrac{8x - 1 = 23 - 4x}{8 \cdot 2 - 1 \mid 23 - 4 \cdot 2}$

$8x + 4x = 23 + 1$

$12x = 24$ $16 - 1 \mid 23 - 8$

$x = \frac{24}{12}$ $15 \mid 15$ TRUE

$x = 2$

35. $2x - 1 = 4 + x$ Check: $\dfrac{2x - 1 = 4 + x}{2 \cdot 5 - 1 \mid 4 + 5}$

$2x - x = 4 + 1$

$x = 5$ $10 - 1 \mid 9$

$9 \mid$ TRUE

37. $6x + 3 = 2x + 11$ Check: $\dfrac{6x + 3 = 2x + 11}{6 \cdot 2 + 3 \mid 2 \cdot 2 + 11}$

$6x - 2x = 11 - 3$

$4x = 8$ $12 + 3 \mid 4 + 11$

$x = \frac{8}{4}$ $15 \mid 15$ TRUE

$x = 2$

39. $5 - 2x = 3x - 7x + 25$

$5 - 2x = -4x + 25$

$4x - 2x = 25 - 5$

$2x = 20$

$x = \frac{20}{2}$

$x = 10$

Check: $\dfrac{5 - 2x = 3x - 7x + 25}{5 - 2 \cdot 10 \mid 3 \cdot 10 - 7 \cdot 10 + 25}$

$5 - 20 \mid 30 - 70 + 25$

$-15 \mid -40 + 25$

$\mid -15$ TRUE

41. $4 + 3x - 6 = 3x + 2 - x$

$3x - 2 = 2x + 2$ Collecting like terms on each side

$3x - 2x = 2 + 2$

$x = 4$

41. (continued)

Check: $\dfrac{4 + 3x - 6 = 3x + 2 - x}{4 + 3 \cdot 4 - 6 \mid 3 \cdot 4 + 2 - 4}$

$4 + 12 - 6 \mid 12 + 2 - 4$

$16 - 6 \mid 14 - 4$

$10 \mid 10$ TRUE

43. $4y - 4 + y + 24 = 6y + 20 - 4y$

$5y + 20 = 2y + 20$

$5y - 2y = 20 - 20$

$3y = 0$

$y = 0$

Check: $\dfrac{4y - 4 + y + 24 = 6y + 20 - 4y}{4 \cdot 0 - 4 + 0 + 24 \mid 6 \cdot 0 + 20 - 4 \cdot 0}$

$0 - 4 + 0 + 24 \mid 0 + 20 - 0$

$20 \mid 20$ TRUE

45. $\frac{7}{2}x + \frac{1}{2}x = 3x + \frac{3}{2} + \frac{5}{2}x$, LCM is 2

$2\left(\frac{7}{2}x + \frac{1}{2}x\right) = 2\left(3x + \frac{3}{2} + \frac{5}{2}x\right)$

$2 \cdot \frac{7}{2}x + 2 \cdot \frac{1}{2}x = 2 \cdot 3x + 2 \cdot \frac{3}{2} + 2 \cdot \frac{5}{2}x$

$7x + x = 6x + 3 + 5x$

$8x = 11x + 3$

$8x - 11x = 3$

$-3x = 3$

$x = \frac{3}{-3}$

$x = -1$

Check: $\dfrac{\frac{7}{2}x + \frac{1}{2}x = 3x + \frac{3}{2} + \frac{5}{2}x}{\frac{7}{2}(-1) + \frac{1}{2}(-1) \mid 3(-1) + \frac{3}{2} + \frac{5}{2}(-1)}$

$-\frac{7}{2} - \frac{1}{2} \mid -3 + \frac{3}{2} - \frac{5}{2}$

$-\frac{8}{2} \mid -\frac{3}{2} - \frac{5}{2}$

$-4 \mid -\frac{8}{2}$

$\mid -4$ TRUE

47. $\frac{2}{3} + \frac{1}{4}t = \frac{1}{3}$, LCM is 12

$12\left(\frac{2}{3} + \frac{1}{4}t\right) = 12 \cdot \frac{1}{3}$

$12 \cdot \frac{2}{3} + 12 \cdot \frac{1}{4}t = 12 \cdot \frac{1}{3}$

$8 + 3t = 4$

$3t = 4 - 8$

$3t = -4$

$t = \frac{-4}{3}$

$t = -\frac{4}{3}$

47. (continued)

Check:
$$\frac{\frac{2}{3} + \frac{1}{4}t = \frac{1}{3}}{\frac{2}{3} + \frac{1}{4}\left(-\frac{4}{3}\right) \;\bigg|\; \frac{1}{3}}$$

$$\frac{2}{3} - \frac{1}{3} \;\bigg|\;$$

$$\frac{1}{3} \quad\text{TRUE}$$

49. $\frac{2}{3} + 3y = 5y - \frac{2}{15}$, LCM is 15

$$15\left[\frac{2}{3} + 3y\right] = 15\left[5y - \frac{2}{15}\right]$$

$$15 \cdot \frac{2}{3} + 15 \cdot 3y = 15 \cdot 5y - 15 \cdot \frac{2}{15}$$

$$10 + 45y = 75y - 2$$

$$10 + 2 = 75y - 45y$$

$$12 = 30y$$

$$\frac{12}{30} = y$$

$$\frac{2}{5} = y$$

Check:
$$\frac{\frac{2}{3} + 3y = 5y - \frac{2}{15}}{\frac{2}{3} + 3 \cdot \frac{2}{5} \;\bigg|\; 5 \cdot \frac{2}{5} - \frac{2}{15}}$$

$$\frac{2}{3} + \frac{6}{5} \;\bigg|\; 2 - \frac{2}{15}$$

$$\frac{10}{15} + \frac{18}{15} \;\bigg|\; \frac{30}{15} - \frac{2}{15}$$

$$\frac{28}{15} \;\bigg|\; \frac{28}{15} \quad\text{TRUE}$$

51. $\frac{5}{3} + \frac{2}{3}x = \frac{25}{12} + \frac{5}{4}x + \frac{3}{4}$, LCM is 12

$$12\left[\frac{5}{3} + \frac{2}{3}x\right] = 12\left[\frac{25}{12} + \frac{5}{4}x + \frac{3}{4}\right]$$

$$12 \cdot \frac{5}{3} + 12 \cdot \frac{2}{3}x = 12 \cdot \frac{25}{12} + 12 \cdot \frac{5}{4}x + 12 \cdot \frac{3}{4}$$

$$20 + 8x = 25 + 15x + 9$$

$$20 + 8x = 15x + 34$$

$$20 - 34 = 15x - 8x$$

$$-14 = 7x$$

$$\frac{-14}{7} = 7x$$

$$-2 = x$$

Check:
$$\frac{\frac{5}{3} + \frac{2}{3}x = \frac{25}{12} + \frac{5}{4}x + \frac{3}{4}}{\frac{5}{3} + \frac{2}{3}(-2) \;\bigg|\; \frac{25}{12} + \frac{5}{4}(-2) + \frac{3}{4}}$$

$$\frac{5}{3} - \frac{4}{3} \;\bigg|\; \frac{25}{12} - \frac{5}{2} + \frac{3}{4}$$

$$\frac{1}{3} \;\bigg|\; \frac{25}{12} - \frac{30}{12} + \frac{9}{12}$$

$$\;\bigg|\; \frac{4}{12}$$

$$\;\bigg|\; \frac{1}{3} \quad\text{TRUE}$$

53. $2.1x + 45.2 = 3.2 - 8.4x$ Greatest number of decimal places is 1

$10(2.1x + 45.2) = 10(3.2 - 8.4)x$ Multiplying by 10 to clear decimals

$$10(2.1x) + 10(45.2) = 10(3.2) - 10(8.4x)$$

$$21x + 452 = 32 - 84x$$

$$21x + 84x = 32 - 452$$

$$105x = -420$$

$$x = \frac{-420}{105}$$

$$x = -4$$

Check:
$$\frac{2.1x + 45.2 = 3.2 - 8.4x}{2.1(-4) + 45.2 \;\bigg|\; 3.2 - 8.4(-4)}$$

$$-8.4 + 45.2 \;\bigg|\; 3.2 + 33.6$$

$$36.8 \;\bigg|\; 36.8 \quad\text{TRUE}$$

55. $1.03 - 0.62x = 0.71 - 0.22x$

Greatest number of decimal places is 2

$$100(1.03 - 0.62x) = 100(0.71 - 0.22x)$$

Multiplying by 100 to clear decimals

$$100(1.03) - 100(0.62x) = 100(0.71) - 100(0.22x)$$

$$103 - 62x = 71 - 22x$$

$$32 = 40x$$

$$\frac{32}{40} = x$$

$$\frac{4}{5} = x, \text{ or}$$

$$0.8 = x$$

Check:
$$\frac{1.03 - 0.62x = 0.71 - 0.22x}{1.03 - 0.62(0.8) \;\bigg|\; 0.71 - 0.22(0.8)}$$

$$1.03 - 0.496 \;\bigg|\; 0.71 - 0.176$$

$$0.534 \;\bigg|\; 0.534 \quad\text{TRUE}$$

57. Do the long division. The answer is negative.

$$3.4_\wedge \overline{\smash)2\,2\,1.0_\wedge 0}$$

with quotient 6.5, $2\,0\,4$, $1\,7\,0$, $1\,7\,0$, 0

$$-22.1 \div 3.4 = -6.5$$

59. Since −15 is to the left of −13 on the number line, −15 is less than −13, so −15 < −13.

61. $\frac{y - 2}{3} = \frac{2 - y}{5}$, LCM is 15

$$15\left[\frac{y - 2}{3}\right] = 15\left[\frac{2 - y}{5}\right]$$

$$5(y - 2) = 3(2 - y)$$

$$5y - 10 = 6 - 3y$$

$$5y + 3y = 6 + 10$$

$$8y = 16$$

$$y = 2$$

63. $\dfrac{5 + 2y}{3} = \dfrac{25}{12} + \dfrac{5y + 3}{4}$,   LCM is 12

$12\left[\dfrac{5 + 2y}{3}\right] = 12\left[\dfrac{25}{12} + \dfrac{5y + 3}{4}\right]$

$4(5 + 2y) = 25 + 3(5y + 3)$

$20 + 8y = 25 + 15y + 9$

$-7y = 14$

$y = -2$

Exercise Set 12.5

1. b + 6, or 6 + b

3. c − 9

5. q + 6, or 6 + q

7. a + b, or b + a

9. y − x

11. w + x, or x + w

13. r + s, or s + r

15. 2x

17. 5t

19. Let x represent the number. Then we have 97%x, or 0.97x.

21. **Familiarize.** Let x = the number.
Translate.

What number, added to, 60 is 112?
$\qquad$ x $\qquad$ + $\quad$ 60 $\;$ = $\;$ 112

Solve. We solve the equation.
$\quad$ x + 60 = 112
$\qquad\quad$ x = 52   Subtracting 60
Check. 52 + 60 = 112, so the number checks.
State. The number is 52.

23. **Familiarize.** Let s = the number of squares your opponent wins. Note that the number of squares won by you and your opponent together must total 64.
Translate. We reword the problem.

Number of   number of
squares   plus   squares your   is 64.
you win   opponent wins
$\quad$ 35 $\qquad$ + $\qquad$ s $\qquad\quad$ = $\;$ 64

Solve. We solve the equation.
$\quad$ 35 + s = 64
$\qquad\quad$ s = 29   Subtracting 35

23. (continued)
Check. Together, you and your opponent would win 35 + 29, or 64 squares. The number checks.
State. Your opponent wins 29 squares.

25. **Familiarize.** Let c = the cost of one 12-oz box of oat flakes. Then four boxes cost 4c.
Translate.

The cost of
four boxes,   was $7.96.
$\quad$ 4c $\qquad$ = $\qquad$ 7.96

Solve. We solve the equation.
$\quad$ 4c = 7.96
$\qquad$ c = 1.99
Check. If one box cost $1.99, then four boxes cost 4($1.99), or $7.96. The result checks.
State. One box cost $1.99.

27. **Familiarize.** Let n = the number.
Translate. We reword the problem.

Six times a number less 18 is 96.
$\quad$ 6 $\;\cdot\;$ n $\qquad$ − $\;$ 18 = 96

Solve. We solve the equation.
$\quad$ 6n − 18 = 96
$\qquad\;$ 6n = 114   Adding 18
$\qquad\;\;$ n = 19   Dividing by 6
Check. Six times 19 is 114. Subtracting 18 from 114, we get 96. This checks.
State. The number is 19.

29. **Familiarize.** Let y = the number.
Translate. We reword the problem.

Two times a number plus 16 is $\frac{2}{5}$ of the number.
$\quad$ 2 $\;\cdot\;$ y $\quad$ + 16 = $\frac{2}{5}\cdot$ $\quad$ y

Solve. We solve the equation.
$\quad$ $2y + 16 = \frac{2}{5}y$
$\quad$ $5(2y + 16) = 5\cdot\frac{2}{5}y$ Clearing the fraction
$\quad$ $10y + 80 = 2y$
$\qquad\;\;$ $80 = -8y$ Subtracting 10y
$\qquad\;$ $-10 = y$ Dividing by −8
Check. We double −10 and get −20. Adding 16, we get −4. Also, $\frac{2}{5}(-10) = -4$. The answer checks.
State. The number is −10.

31. <u>Familiarize</u>. First draw a picture.

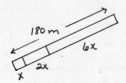

We use x for the first length, 2x for the second length, and 3·2x, or 6x, for the third length.

<u>Translate</u>. The lengths of the three pieces add up to 180 m. This gives us the equation.

Length   Length   Length
of 1st plus of 2nd plus of 3rd is 180
piece   piece   piece

  x   +   2x   +   6x   = 180

<u>Solve</u>. We solve the equation.

$$x + 2x + 6x = 180$$
$$9x = 180$$
$$x = 20$$

<u>Check</u>. If the first piece is 20 m long, then the second is 2·20 m, or 40 m and the third is 6·20 m, or 120 m. The lengths of these pieces add up to 180 m (20 + 40 + 120 = 180). This checks.

<u>State</u>. The first piece measures 20 m. The second measures 40 m, and the third measures 120 m.

33. <u>Familiarize</u>. We draw a picture. Let w = the width of the rectangle. Then w + 60 = the length.

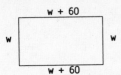

The perimeter of a rectangle is the sum of the lengths of the sides. The area is the product of the length and the width.

<u>Translate</u>. We use the definition of perimeter to write an equation that will allow us to find the width and length.

Width + Width + Length + Length = Perimeter
  w + w + (w + 60) + (w + 60) =   520

To find the area we will compute the product of the length and width, or (w + 60)w.

<u>Solve</u>. We solve the equation.

$$w + w + (w + 60) + (w + 60) = 520$$
$$4w + 120 = 520$$
$$4w = 400$$
$$w = 100$$

If w = 100, then w + 60 = 100 + 60 = 160, and the area is 160(100) = 16,000.

<u>Check</u>. The length is 60 ft more than the width. The perimeter is 100 + 100 + 160 + 160 = 520 ft. This checks. To check the area we recheck the computation. This also checks.

<u>State</u>. The width is 100 ft, the length is 160 ft, and the area is 16,000 ft².

35. <u>Familiarize</u>. We draw a picture. Let ℓ = the length of the paper. Then ℓ - 6.3 = the width.

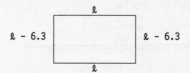

The perimeter is the sum of the lengths of the sides.

<u>Translate</u>. We use the definition of perimeter to write an equation.

  Width +   Width   + Length + Length is 99.
(ℓ - 6.3) + (ℓ - 6.3) +   ℓ   +   ℓ   = 99

<u>Solve</u>. We solve the equation.

$$(ℓ - 6.3) + (ℓ - 6.3) + ℓ + ℓ = 99$$
$$4ℓ - 12.6 = 99$$
$$4ℓ = 111.6$$
$$ℓ = 27.9$$

Then ℓ - 6.3 = 21.6.

<u>Check</u>. The width, 21.6 cm, is 6.3 cm less than the length, 27.9 cm. The perimeter is 21.6 cm + 21.6 cm + 27.9 cm + 27.9 cm, or 99 cm. This checks.

<u>State</u>. The length is 27.9 cm, and the width is 21.6 cm.

37. <u>Familiarize</u>. The total cost is the daily charge plus the mileage charge. The mileage charge is the cost per mile times the number of miles driven. Let m = the number of miles that can be driven for $80.

<u>Translate</u>.

Daily plus Cost per times Number of is Amount
rate   mile   miles driven
34.95   +   0.10   ·   m   = 80

<u>Solve</u>. We solve the equation.

$$34.95 + 0.10m = 80$$
$$100(34.95 + 0.10m) = 100(80) \quad \text{Clearing the decimals}$$
$$3495 + 10m = 8000$$
$$10m = 4505$$
$$m = 450.5$$

<u>Check</u>. The mileage cost is found by multiplying 450.5 by $0.10 obtaining $45.05. Then we add $45.05 to $34.95, the daily rate, and get $80.

<u>State</u>. The businessperson can drive 450.5 mi on the car-rental allotment.

39. <u>Familiarize</u>. Let s = one score. Then four
    score = 4s and four score and seven = 4s + 7.

    <u>Translate</u>. We reword.

    1776 plus four score is 1863.
                and seven

    1776   +   (4s + 7)   =   1863

    <u>Solve</u>. We solve the equation.

    $$1776 + (4s + 7) = 1863$$
    $$4s + 1783 = 1863$$
    $$4s = 80$$
    $$s = 20$$

    <u>Check</u>. If a score is 20 years, then four score
    and seven represents 87 years. Adding 87 to
    1776 we get 1863. This checks.

    <u>State</u>. A score is 20.

41. <u>Familiarize</u>. Let ℓ = the length of the rectangle.
    Then the width is $\frac{3}{4}\ell$. When the length and width
    are each increased by 2 cm they become ℓ + 2 and
    $\frac{3}{4}\ell$ + 2, respectively. We will use the formula for
    perimeter, $2\ell + 2w = P$.

    <u>Translate</u>. We substitute ℓ + 2 for ℓ, $\frac{3}{4}\ell$ + 2
    for w, and 50 for P in the formula.

    $$2(\ell + 2) + 2\left[\frac{3}{4}\ell + 2\right] = 50$$

    <u>Solve</u>. We solve the equation.

    $$2(\ell + 2) + 2\left[\frac{3}{4}\ell + 2\right] = 50$$
    $$2\ell + 4 + \frac{3}{2}\ell + 4 = 50$$
    $$2\left[2\ell + 4 + \frac{3}{2}\ell + 4\right] = 2(50) \quad \text{Clearing the fraction}$$
    $$4\ell + 8 + 3\ell + 8 = 100$$
    $$7\ell + 16 = 100$$
    $$7\ell = 84$$
    $$\ell = 12$$

    Possible dimensions are ℓ = 12 cm and
    w = $\frac{3}{4} \cdot$ 12 cm = 9 cm.

    <u>Check</u>. If the length is 12 cm and the width is
    9 cm, then they become 12 + 2, or 14 cm, and
    9 + 2, or 11 cm, respectively, when they are each
    increased by 2 cm. The perimeter becomes
    $2 \cdot 14 + 2 \cdot 11$, or 28 + 22, or 50 cm. This checks.

    <u>State</u>. The length is 12 cm, and the width is
    9 cm.

43. <u>Familiarize</u>. Let x = number of half dollars.
    Then 2x = number of quarters,
          4x = number of dimes ($2 \cdot 2x = 4x$), and
          12x = number of nickels ($3 \cdot 4x = 12x$).
    The value of x half dollars is 0.50(x).
    The value of 2x quarters is 0.25(2x).
    The value of 4x dimes is 0.10(4x).
    The value of 12x nickels is 0.05(12x).

    <u>Translate</u>. The total value is $10.

    $$0.50(x) + 0.25(2x) + 0.10(4x) + 0.05(12x) = 10$$

    <u>Solve</u>.

    $$0.50(x) + 0.25(2x) + 0.10(4x) + 0.05(12x) = 10$$
    $$0.5x + 0.5x + 0.4x + 0.6x = 10$$
    $$2x = 10$$
    $$x = 5$$

    Possible answers for the number of each coin:

    Half dollars = x = 5
    Quarters = 2x = 2·5 = 10
    Dimes = 4x = 4·5 = 20
    Nickels = 12x = 12·5 = 60

    <u>Check</u>. The value of

    5 half dollars = $2.50
    10 quarters   = 2.50
    20 dimes   = 2.00
    60 nickels   = 3.00

    The total value is $10. The numbers check.

    <u>State</u>. The storekeeper got 5 half dollars, 10
    quarters, 20 dimes, and 60 nickels.